NATIONAL ACADEMIES *Sciences Engineering Medicine*

NATIONAL ACADEMIES PRESS
Washington, DC

Tipping Points, Cascading Impacts, and Interacting Risks in the Earth System

Katrina Hui, Hugh Walpole, and Margo Corum, *Rapporteurs*

Committee on Tipping Points, Cascading Impacts, and Interacting Risks in the Earth System: A Workshop

Board on Atmospheric Sciences and Climate

Board on Earth Sciences and Resources

Division on Earth and Life Studies

Proceedings of a Workshop

NATIONAL ACADEMIES PRESS 500 Fifth Street, NW, Washington, DC 20001

This activity was supported by Grant No. 2022844 between the National Academy of Sciences and the National Science Foundation. Any opinions, findings, conclusions, or recommendations expressed in this publication do not necessarily reflect the views of any organization or agency that provided support for the project.

International Standard Book Number-13: 978-0-309-70134-1
International Standard Book Number-10: 0-309-70134-1
Digital Object Identifier: https://doi.org/10.17226/26925

This publication is available from the National Academies Press, 500 Fifth Street, NW, Keck 360, Washington, DC 20001; (800) 624-6242 or (202) 334-3313; http://www.nap.edu.

Copyright 2024 by the National Academy of Sciences. National Academies of Sciences, Engineering, and Medicine and National Academies Press and the graphical logos for each are all trademarks of the National Academy of Sciences. All rights reserved.

Printed in the United States of America.

Suggested citation: National Academies of Sciences, Engineering, and Medicine. 2024. *Tipping Points, Cascading Impacts, and Interacting Risks in the Earth System: Proceedings of a Workshop.* Washington, DC: The National Academies Press. https://doi.org/10.17226/26925.

The **National Academy of Sciences** was established in 1863 by an Act of Congress, signed by President Lincoln, as a private, nongovernmental institution to advise the nation on issues related to science and technology. Members are elected by their peers for outstanding contributions to research. Dr. Marcia McNutt is president.

The **National Academy of Engineering** was established in 1964 under the charter of the National Academy of Sciences to bring the practices of engineering to advising the nation. Members are elected by their peers for extraordinary contributions to engineering. Dr. John L. Anderson is president.

The **National Academy of Medicine** (formerly the Institute of Medicine) was established in 1970 under the charter of the National Academy of Sciences to advise the nation on medical and health issues. Members are elected by their peers for distinguished contributions to medicine and health. Dr. Victor J. Dzau is president.

The three Academies work together as the **National Academies of Sciences, Engineering, and Medicine** to provide independent, objective analysis and advice to the nation and conduct other activities to solve complex problems and inform public policy decisions. The National Academies also encourage education and research, recognize outstanding contributions to knowledge, and increase public understanding in matters of science, engineering, and medicine.

Learn more about the National Academies of Sciences, Engineering, and Medicine at **www.nationalacademies.org**.

Consensus Study Reports published by the National Academies of Sciences, Engineering, and Medicine document the evidence-based consensus on the study's statement of task by an authoring committee of experts. Reports typically include findings, conclusions, and recommendations based on information gathered by the committee and the committee's deliberations. Each report has been subjected to a rigorous and independent peer-review process and it represents the position of the National Academies on the statement of task.

Proceedings published by the National Academies of Sciences, Engineering, and Medicine chronicle the presentations and discussions at a workshop, symposium, or other event convened by the National Academies. The statements and opinions contained in proceedings are those of the participants and are not endorsed by other participants, the planning committee, or the National Academies.

Rapid Expert Consultations published by the National Academies of Sciences, Engineering, and Medicine are authored by subject-matter experts on narrowly focused topics that can be supported by a body of evidence. The discussions contained in rapid expert consultations are considered those of the authors and do not contain policy recommendations. Rapid expert consultations are reviewed by the institution before release.

For information about other products and activities of the National Academies, please visit www.nationalacademies.org/about/whatwedo.

WORKSHOP PLANNING COMMITTEE ON TIPPING POINTS, CASCADING IMPACTS, AND INTERACTING RISKS IN THE EARTH SYSTEM

KRISTEN ST. JOHN (*Chair*), James Madison University
AMIR AGHAKOUCHAK, University of California, Irvine
KATHARINE V. CASHMAN, University of Oregon
SIMON DIETZ, London School of Economics and Political Science
TIMOTHY M. LENTON, University of Exeter
DOROTHY J. MERRITTS, Franklin and Marshall College
MICHAEL SCHOON, Arizona State University

National Academies of Sciences, Engineering, and Medicine Staff

MARGO CORUM, Study Director (*until December 2023*)
KATRINA HUI, Associate Program Officer (*from February 2024*)
PATRICIA RAZAFINDRAMBININA, Associate Program Officer (*until April 2023*)
HUGH WALPOLE, Associate Program Officer (*until July 2024*)
MAYA FREY, Senior Program Assistant (*from April 2024*)
SABAH RANA, Senior Program Assistant (*until May 2023*)

Reviewers

This Proceedings of a Workshop on Tipping Points, Cascading Impacts, and Interacting Risks in the Earth System was reviewed in draft form by individuals chosen for their diverse perspectives and technical expertise. The purpose of this independent review is to provide candid and critical comments that will assist the National Academies of Sciences, Engineering, and Medicine in making each published proceedings as sound as possible and to ensure that it meets the institutional standards for quality, objectivity, evidence, and responsiveness to the charge. The review comments and draft manuscript remain confidential to protect the integrity of the process.

We thank the following individuals for their review of this proceedings:

LEE KUMP (NAS), Pennsylvania State University
MICHAEL SCHOON, Arizona State University

Although the reviewers listed above provided many constructive comments and suggestions, they were not asked to endorse the content of the proceedings, nor did they see the final draft before its release. The review of this proceedings was overseen by **JAMES HURRELL,** Colorado State University. He was responsible for making certain that an independent examination of this proceedings was carried out in accordance with standards of the National Academies and that all review comments were carefully considered. Responsibility for the final content rests entirely with the rapporteur(s) and the National Academies.

We also thank staff member Samantha Koretsky for reading and providing helpful comments on this manuscript.

Contents

1 **INTRODUCTION** 1
   Workshop Description, 2
   Welcome from the National Science Foundation, 2
   Background and Motivation for Workshop, 3
   Overview on Tipping Points and Expert Response, 5
   Moderated Q&A Discussion with Expert Speakers, 11

2 **HISTORICAL ANALYSIS OF PAST BIOGEOPHYSICAL AND SOCIAL TIPPING POINTS** 14
   Economic Perspective on Social Tipping Points Relating to Climate Change, 14
   Historical Example of a Tipping Point: The Dust Bowl 1930s—The Perfect Disaster, 15
   A Simplified Model of Biogeophysical Tipping Points: Lovelock's Daisyworld, 17
   Breakout Group Discussions on Biogeophysical and Social Tipping Points, 18

3 **REGIONAL PERSPECTIVES ON CLIMATE TIPPING POINTS AND CASCADING RISK** 21
   Coastal United States: Tipping Points in Future Tropical Pacific Island Sustainability, 21
   American West: Summer Water in the West—Climate Tipping Points and Cascading Impacts, 23
   Arctic Perspectives, 26
   Great Plains: Tipping Points and Cascading Impacts in Nebraska, 30
   Q&A Discussion with Regional Perspectives: Overarching Themes, 33
   Breakout Group Discussions on Regional Perspectives on Climate Tipping Points and Cascading Impacts, 36

4 **EXAMPLES OF INTERDISCIPLINARY RESEARCH PRIORITIES AND OPPORTUNITIES** 39
   Insights from Indigenous Knowledge, 39
   Cascading Impacts and Environmental Justice, 40
   Moderated Q&A Discussion with Experts, 42
   Breakout Group Discussions on Interdisciplinary Research Topics and Opportunities, 45

5 **WORKSHOP SYNTHESIS, THEMES, AND CLOSING THOUGHTS** 47
   Synthesis of Workshop Discussions, 47
   High-Level Workshop Themes, 48
   Closing Thoughts, 51

REFERENCES 52

APPENDIXES

A   WORKSHOP AGENDA 55

B   STATEMENT OF TASK 59

C   PLANNING COMMITTEE BIOGRAPHIES 60

1
Introduction

Earth is a complex system, with myriad interactions and feedbacks among the atmosphere, hydrosphere, geosphere, cryosphere, and biosphere, as well as the people, institutions, and technologies that respond to and influence these dynamics. Many pressing societal concerns involve processes that interact in dynamic and often nonlinear ways. A systems-based approach to scientific research can yield understanding across all aspects of Earth's interconnected processes. Earth systems studies rest on careful analyses of physical, chemical, biological, and behavioral processes and their interactions and feedbacks; these studies are supported by observational platforms, laboratory and modeling facilities, a skilled scientific workforce, and the cyberinfrastructure that connects them to enable new breakthroughs.

In September 2021, the National Academies released the consensus study report, *Next Generation Earth Systems Science at the National Science Foundation* (NASEM, 2022), which identified key characteristics for next generation Earth systems science. To continue a robust post-release engagement aligned with the overall study's goals, this workshop was organized to explore the approaches to Earth systems science through the lens of a critical topic—tipping points—and to cultivate cross-disciplinary collaborations and prime new research communities to discuss examples of research needs in a broader context of Earth systems science. This rich topic emphasizes the importance of research on complex interconnections and feedbacks between natural and social processes, and illuminates how the next generation of Earth systems science can help develop different pathways to address the challenges society faces.

Here, *tipping points* refers to the concept that climate change could drive elements of Earth's environment past a threshold, leading to abrupt, irreversible shifts with dangerous consequences. The risks of tipping points in important aspects and processes in the physical and ecological systems have long been explored, including those related to runaway loss of polar ice sheets and resulting sea level rise; rapid release of greenhouse gases from thawing permafrost; major disruptions to ocean circulation and dynamics; and rapid drying of forest ecosystems that alters regional hydrology, fire regimes, and carbon sources/sinks. Recent interest in the concept of "social tipping points" has emerged, where major societal stresses such as food, energy, or water shortages, devastation from extreme weather, or pandemics and other health risks can accumulate to a point that society is pushed into radically new dynamics (e.g., conflict, mass migration, major demographic shifts), and adds a layer of interconnectedness and complexity.

Tipping points can only be fully understood when their connections to related cascading impacts and interacting risks are known. *Cascading impacts* are a sequence of events where abrupt changes in one component lead to abrupt changes in other components. *Interacting risks* are best conceptualized as tipping elements whose interaction on a global scale could have stabilizing or destabilizing effects, thereby increasing or decreasing the probability of cascading impacts.

To address these challenges, this workshop convened experts across social, natural, computational, and engineering sciences to engage in transdisciplinary dialogues about advancing understanding, prediction, and preparation for tipping points, cascading impacts, and interacting risks in the Earth system—drawing connections to the research insights conveyed in the *Next Generation Earth Systems Science at the National Science Foundation* (NASEM, 2022) report.

WORKSHOP DESCRIPTION

The National Academies of Sciences, Engineering, and Medicine convened a workshop on January 17–19, 2023, in Washington, DC, and online, to consider the state of understanding and integrated approaches for climate tipping points, cascading impacts, and interacting risks in the Earth system. A planning committee was chosen to plan the workshop.[1]

The workshop focused on applying Earth systems science approaches to explore understanding, prediction, and preparation for both environmental and social tipping points in the Earth system and to foster connections among the transdisciplinary research community. The workshop explored historical examples of past biogeophysical and social tipping points and a range of regional U.S. perspectives on climate tipping points and cascading impacts through a mix of presentations, panel discussions, and interactive breakout sessions. Finally, the workshop culminated with group and breakout discussions, during which workshop participants focused on discussing aims and opportunities for interdisciplinary research. See Appendix A for the workshop agenda, Appendix B for the Statement of Task, and Appendix C for Planning Committee biographies.

WELCOME FROM THE NATIONAL SCIENCE FOUNDATION

The workshop opened with brief welcomes from sponsors Dr. Anjuli Bamzai and Dr. Candace Major from the National Science Foundation (NSF). They noted the importance of the consensus study that produced *Next Generation Earth Systems Science at the National Science Foundation* (NASEM, 2022) and this workshop in gathering

---

[1] These proceedings have been prepared by the workshop rapporteur as a factual summary of what occurred at the workshop. The planning committee's role was limited to planning and convening the workshop. The views contained in the proceedings are those of individual workshop participants and do not necessarily represent the views of all workshop participants, the planning committee, or the National Academies of Sciences, Engineering, and Medicine.

Introduction 3

community input on emerging research questions that have significant societal implications in order to inform the work of NSF, particularly its geosciences directorates.

BACKGROUND AND MOTIVATION FOR WORKSHOP

Following the welcome from NSF, the workshop planning committee chair, Dr. Kristen St. John, James Madison University, introduced workshop goals and results from an informal pre-workshop poll. She then reviewed the study premise, vision, and key characteristics and recommendations from the report to connect to the workshop framing. St. John explained that the study called for an integrated approach between natural and social sciences and processes to understand the Earth's systems and capacity for sustaining life now and in the future, and as such, that this was also a key component of the workshop. St. John then presented the six key characteristics of an integrated approach to next generation Earth systems science, as identified in the report, which are to:

1. **Advance** both curiosity-driven and use-inspired basic **research** on the Earth's systems across spatial, temporal, and social organization scales.
2. **Facilitate convergence** of social, natural, computational, and engineering sciences to advance science and inform solutions to Earth systems–related problems.
3. Ensure **diverse, inclusive, equitable, and just** approaches to Earth systems science.
4. **Prioritize engagement** and **partnerships** with diverse stakeholders to benefit society and address Earth systems–related problems at community, state, national, and international scales.
5. Use **observational, computational,** and **modeling capabilities** synergistically to accelerate discovery and convergence.
6. **Educate** and **support a workforce** with the skills and knowledge to effectively identify, conduct, and communicate Earth systems science.

She drew connections between these characteristics and the framing and guiding questions of the workshop, including advancing curiosity- and use-inspired research across a range of scales; facilitating convergence across a wide range of disciplines; and ensuring diversity, inclusivity, and equity are incorporated into Earth systems science research approaches and outcomes. Particularly, she noted, this workshop topic on tipping points, cascading impacts, and interacting risks in the Earth system benefits from including broad transdisciplinary[2] perspectives to address contemporary societal challenges through open exchanges of information, and shared learning, as described in the consensus study report.

---

[2] As defined in *Next Generation of Earth System Science at the National Science Foundation* (NASEM, 2022), transdisciplinary research transcends disciplinary approaches to create a new framework and approach fostering collaboration among researchers and communities across a range of disciplines to address shared societal needs and challenges.

St. John guided the participants deeper into the workshop tasks, explaining that in addition to incorporating the report's key characteristics, the committee was tasked to apply Earth systems science approaches to explore understanding, prediction, and preparation for Earth system tipping points, and to support cultivating connections among transdisciplinary research communities to address questions and brainstorm potential opportunities, barriers, and strategies. She noted that this workshop fits best into the first two of the four phases of the Hall model of transdisciplinary team-based research, which are developmental, conceptual, implementation, and translational (Hall et al., 2012). This workshop has a goal of understanding tipping points with a team-type of a community or network of experts from a range of disciplines and a process of working toward a shared goal on the workshop topic. To foster completion of these phases and bridge-building across the broad range of disciplines, she emphasized the workshop would encourage idea sharing on these key themes through a combination of plenary and interactive breakout sessions.

St. John then briefly reviewed the workshop agenda, which included an overview on tipping points; historical analyses and perspectives on tipping points, cascading effects, and interacting risks over different time periods and regions; and discussions of opportunities and research questions for transdisciplinary research. She discussed the results from an informal, three-question, pre-workshop poll, which informed development of workshop content and structure. The first question asked poll participants to "briefly summarize what tipping points in the Earth's system mean to you." The results, St. John explained, revealed two types of responses. The first theme considered tipping points to be more technically defined as something with rapid nonlinear changes to a different state that is often irreversible. The second theme defined tipping points around looser narratives regarding impacts, e.g., social or economic, resulting from crossing tipping points, such as the destabilization of Earth system components like ecosystems, and the resulting detrimental implications to human civilization. The second question asked poll participants "What are key outstanding research questions on natural (physical, chemical, biological, and social) tipping points, their interacting risks, and their cascading impacts?" St. John noted that five main topics arose from the responses:

1. Conflict and social response
2. Better evidence, reduced uncertainty, and model improvements and related early warning systems
3. Policy responses, institutional responses, and resiliency management
4. Interconnections between tipping points
5. Communicating the issues

St. John summarized the results of the third poll question—"What are the major barriers and opportunities to accelerate progress to advance these areas of research?" Barriers identified by workshop poll participants listed data availability, particularly for social data; missing processes due to limited model resolution and system complexity; and challenges

Introduction 5

to performing interdisciplinary[3] research as barriers. Opportunities lay in integration of models, access to big data and improved Earth observations, and cultivation of interdisciplinary connections. St. John noted that data and interdisciplinary research spanned both barriers and opportunities.

St. John then described how the poll participants ranked topics of interest under three categories: (1) research topics, (2) geographic U.S. regions, and (3) strategies for moving forward—which helped the committee select topics and structure the workshop. She concluded her introduction by emphasizing the importance of developing strategies to research tipping points that support inclusive and equitable participation and engagement. St. John noted that, in alignment with this aim, the workshop and poll participants are experts and stakeholders across a range of disciplines (e.g., natural science, social science, and other technical fields), organizations, and career stages.

OVERVIEW ON TIPPING POINTS AND EXPERT RESPONSE

Workshop committee member Dr. Tim Lenton, University of Exeter, provided a deeper overview of tipping points. This overview was followed by two expert responses, with perspectives on climate tipping points provided by Dr. Jeffrey N. Rubin and Dr. Robert Kopp, Rutgers University.

Tipping Points Overview

Dr. Lenton began his introduction with a description of a simple model of a system with two alternative stable states. The system starts in one stable state, and the model allows for variations. With time, as one state becomes less stable, the model forces the system past a tipping point, where reinforcing feedbacks take over and lead to self-propelling, irreversible change to the other state. Lenton noted that lower stability amplifies the effects of the same amount of variance of the system, which could trigger early warning signals for tipping points through changes in the system autocorrelation.

Lenton described how tipping points exist across the entire range of scales, from Earth, to climate, and to ecological systems, and can be positive,[4] negative,[5] or somewhere in between. Using a schematic of examples plotted against spatial scale and timescale (Figure 1-1), Lenton provided several examples of tipping points from Earth's history (Lenton, 2020). Hundreds to tens of millions of years ago, profound global tipping point

---

[3] As defined in *Next Generation of Earth System Science at the National Science Foundation* (NASEM, 2022), interdisciplinary research integrates information, data, techniques, tools, perspectives, concepts, and/or theories from two or more disciplines focused on a complex question, problem, topic, or theme.
[4] Lenton defines a positive tipping point as one that increases sustainability and social justice, such as a tipping point leading to a significant reduction in cumulative greenhouse gas emissions.
[5] Lenton defines a negative tipping point as one that leads to net harm to a significant number of people, noting that most climate tipping points are negative.

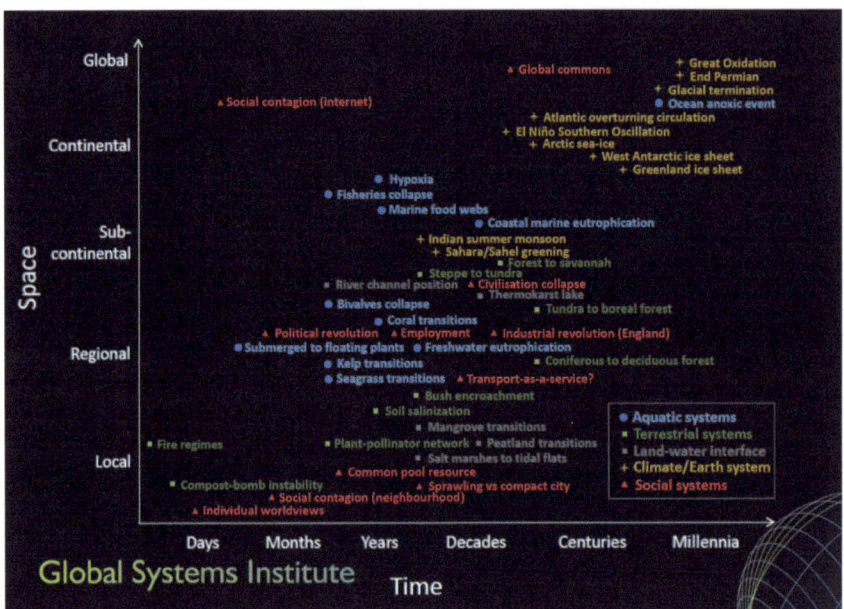

FIGURE 1-1 Examples of Earth's tipping points plotted by spatial and temporal scale.
SOURCE: Lenton (2020).

events occurred, such as the Great Oxidation Event, when the Earth's atmosphere abruptly and irreversibly switched from anoxic to oxidizing. Over this timescale and up to more recent events (within the past few millions of years), Lenton explained that the Earth oscillated between various global climate states, sometimes with abrupt climate change transitions in between, as evidenced by temperature proxy records within Greenland ice cores. Lenton explained that these abrupt climate changes have global effects throughout both hemispheres, with past changes correlating to the rise and fall of past civilizations through the abrupt switching on and off of the tropical monsoon systems.

Lenton described a recent study (Armstrong McKay et al., 2022) that synthesized around 230 studies on tipping points to identify tipping elements, defined as parts of the climate system that could be tipped within this century. He explained that the researchers categorized the tipping elements into (1) global core climate tipping elements (examples in Figure 1-2a), tipping elements that when tipped would be felt throughout the entire climate system, and (2) regional impact tipping elements (examples in Figure 1-2b), which when tipped may not be felt in the whole climate system but could still have significant regional impacts. These tipping elements were spatially identified over particular regions that display evidence of being tipped and experiencing abrupt shifts in the past. Furthermore, Lenton explained that by using observational data, model projections, and offline models, the researchers found that several of these abrupt shifts of the tipping elements occur at relatively low levels of global warming above pre-industrial conditions.

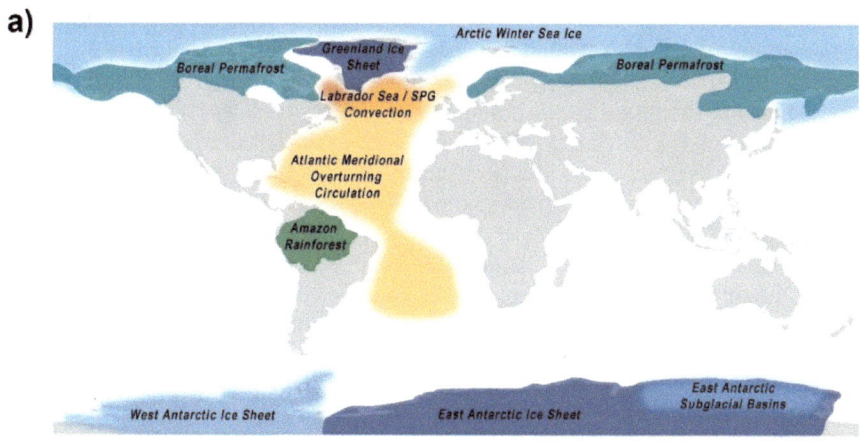

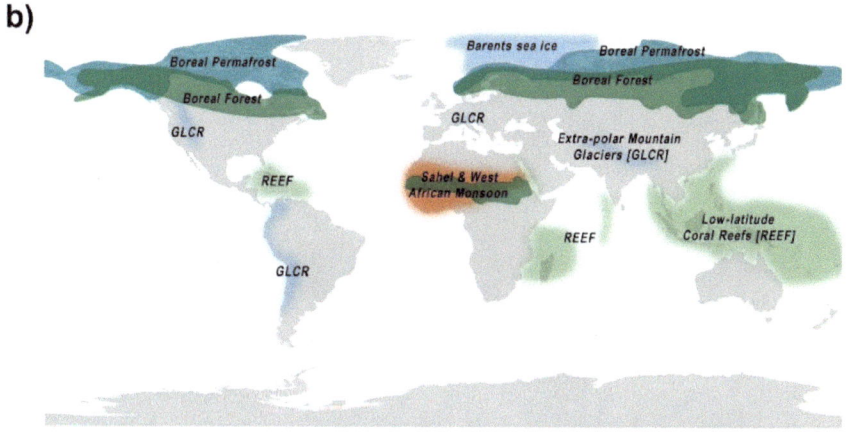

FIGURE 1-2 (a) Global core climate tipping elements and (b) regional impact climate tipping elements.
SOURCE: Armstrong McKay et al. (2022).

Lenton provided an example of a tipping point that occurs within a model, where 2°C of global mean warming results in a tipping point that connects large changes across Antarctic ice sheet loss, Arctic sea ice loss, changes in the Atlantic overturning circulation, and droughts in the Amazon. Changes in the mean state of systems and more persistent and larger amplitude fluctuations in the system can provide insights and early warning signs into where and when tipping points are being approached. After surveying the studies, Armstrong McKay et al. (2022) identified five systems that are potentially at risk of tipping at the current level of approximately 1.2°C global warming: Labrador Sea subpolar gyre,

Greenland ice sheet, West Antarctic ice sheet, low-latitude coral reefs, and portions of the boreal permafrost. Lenton explained that each of the five systems has its own characteristic timescale over which the tipping is expected to unfold. Lenton warned that under current policy, society may continue to expect global warming to reach levels of 2.7°C, which may place approximately 13 of McKay et al.'s identified tipping elements at risk. Furthermore, Lenton noted that because of the interconnected nature of the Earth system, the occurrence of one tipping point may increase the likelihood of another tipping point occurring, creating cascading effects (Figure 1-3).

Dr. Lenton next drew attention to the relative scarcity of research focused on the impacts of passing tipping points. Using the example of collapse of the Atlantic Meridional Overturning Circulation (AMOC) with 2.5°C global warming, Lenton noted that this tipping point would result in severe global temperature and precipitation changes that would impact regional and global agriculture and water resources, which may have additive and/or multiplicative implications worldwide (Figure 1-4). Lenton emphasized that climate tipping points may likely interact with tipping points on other scales of system, resulting in cascading tipping points across and between the climate, ecological, and social systems.

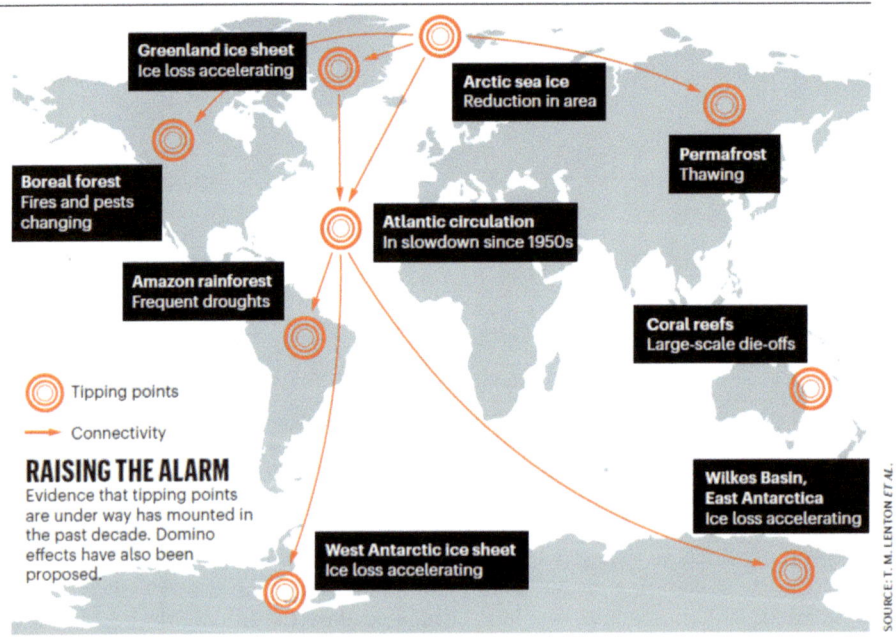

FIGURE 1-3 Tipping points may already be under way and are interconnected, potentially leading to cascading impacts.
SOURCE: Lenton et al. (2019).

Introduction

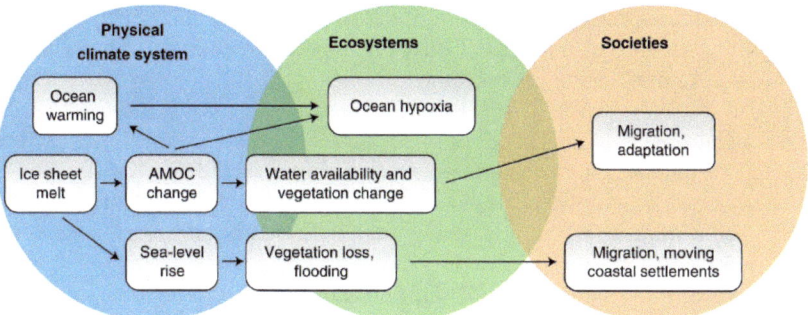

FIGURE 1-4 How abrupt change can cascade across systems during climate warming periods. SOURCE: Brovkin et al. (2021).

Lenton concluded his presentation by looking forward, noting that transdisciplinary work is still needed to better understand the implications of fundamental tipping points within the climate and environment on tipping points in other scales of systems. Lenton noted that 1.5°C of global warming can potentially trigger several climate tipping points, which have early warning signs that have been observed recently. He underscored that these climate tipping points will add to the effects of global warming, potentially resulting in existential risks. Lenton stressed the need for more work to understand fundamental aspects of tipping points to better assess and address their associated risks across scales and systems.

Expert Response on Local-Level Adaptation Actions

Dr. Jeffrey Rubin followed Lenton's overview with an expert response. Relating to the potential impacts of climate change and tipping points, he emphasized the need to consider the local level when addressing the effects of climate change, especially in terms of adaptation actions. Rubin noted that the concept of the "inverse care law" is useful when considering choices for effective adaptation. The inverse care law highlights that critical services such as medical care, social services, and other basic needs tend to be less accessible to populations most in need of them (Hart, 1971).

Rubin discussed four examples of local adaptation choices in the United States and their unique challenges:

- Managed retreat strategies in New Jersey and accommodation approaches in New York City following Hurricane Sandy.
- Puerto Rico's adaptation actions to address environmental challenges despite limited resources.
- Large-scale managed retreat efforts and the related social and political challenges experienced in the U.S. Gulf Coast region, including Texas, Louisiana, and Florida.

- Oregon's response and recovery from the wildfires in 2020, which had a disproportionate and lasting impact on vulnerable communities, such as through the destruction of affordable housing.

To conclude, Rubin underlined the importance of focusing on society's most vulnerable populations in local climate change adaptation efforts, where the challenges and nature are inherently complex, multifaceted, long-term, and expensive. Rubin urged scientists to address the needs of those who are least able to independently navigate these issues.

Expert Response on the Use of "Tipping Points" and Relevance to Policy

Next, Dr. Robert Kopp, Rutgers University, provided an alternate perspective on the concept of tipping points in climate science. Acknowledging the fundamental impact of Lenton's and other's climate tipping points work on scientific understanding of the Earth system, Kopp noted that the term *tipping points* was initially a metaphor borrowed from sociological work, primarily adopted into the climate literature for communication purposes. First used academically in the 1950s in sociological literature on housing segregation, the term was especially popularized by Malcolm Gladwell in *The Tipping Point: How Little Things Can Make a Big Difference* in the late 1990s (Gladwell, 2000). Kopp pondered whether the term brings more clarity or confusion when communicating with the broader public on societally relevant climate risks. Elaborating on this point, he posited that the framing of tipping points might be influenced by communication choices rather than the science itself.

Kopp expressed skepticism about the effects of tipping points discourse on public understanding and referred to limited literature attempting to investigate empirically the effects of "tipping points" on this discourse. He cited a recent study that found that nonlinear portrayals of climate risk, such as those invoked by using the term *tipping points*, may not significantly impact perceptions of the potential for climate catastrophes or the controllability of consequences (Formanski et al., 2022).

Furthermore, Kopp questioned the relevance of global-scale climate tipping points for mitigation policies and local adaptation decisions, and whether they should have a substantial impact on these policies and decisions, as opposed to considering regional ecosystems. He emphasized the broad uncertainty surrounding global-scale climate tipping points and suggested that, although useful because of differences in spatial and timescales, study of the integrated Earth system may not directly influence the mitigation strategies happening now.

Kopp reiterated that despite his skepticism, he still believes that studying large-scale climate tipping points and their potential impacts are important but urged awareness of the communicative effects of the "tipping point" metaphor. He suggested that the tipping points analogy may be better suited for deliberations about how changes in the Earth system may affect human systems. As such, Kopp proposed that a focus on the interaction between the Earth system and human systems, including social tipping points, might be more urgent and relevant for mitigation and adaptation decision making. He added that abrupt nonlinear changes in social systems can occur with or without necessary reliance on physical tipping points in the Earth system and climate. To conclude, Kopp argued for a critical examination of the term *tipping points* and suggested that understanding the linear and nonlinear relationships between climate changes and human system responses and how those responses interact should be a research priority for informing mitigation and adaptation strategies.

MODERATED Q&A DISCUSSION WITH EXPERT SPEAKERS

Following the presentations, Kopp, Lenton, and Rubin took questions from workshop participants. The next sections recount some of the key themes that emerged from the discussion.

Preventing Negative Tipping Points versus Achieving Positive Tipping Points

Regarding the feasibility of preventing negative tipping points and potentially achieving positive tipping points, Lenton emphasized the importance of studying positive tipping points. He defined these as self-reinforcing, self-propelling changes, such as transitions to renewable energy, because they provide an empowering sense of agency in addressing climate change. After reiterating his consideration of tipping points for social systems, Kopp agreed with Lenton regarding the potential to achieve positive tipping points and the benefits of framing environmental governance using this concept. Kopp further suggested that prevention of negative tipping points and achievement of positive ones are interconnected and that catalyzing positive societal tipping points is perhaps a prerequisite for preventing negative Earth system tipping points.

Study of Past Global Warming Events

When asked about the importance of studying historical global warming events to understand the potential for future tipping points, Lenton noted the usefulness of studying past events such as the Paleocene-Eocene Thermal Maximum to understand potential feedback processes and the fundamental and nonlinear aspects of the behavior of Earth's system. Lenton cautioned, however, that scientists should be careful about carrying insights about past events to the present, because the background climates for these events were drastically different. Kopp added that the paleorecord of these past global warming events serves as a source for natural experiments that take the climate system out of its current state to test and improve climate models.

Timescales of Societal System versus Earth System Tipping Points

All three speakers agreed that societal tipping points generally happen more rapidly than Earth system tipping points, although sometimes they may overlap. Lenton and Kopp noted examples, such as social systems, population growth, resource use, technological advancement and adoption, and land use change, that may exhibit tipping behavior on faster timescales. Rubin stressed the challenges of framing the geological timescales of Earth system tipping points for policy and decision-making, which generally focus on timescales on the order of 10 years. Kopp added that given the drastic difference in timescales associated with social versus Earth system tipping points and the relevant decadal timescale for policymakers, he would prioritize exploration of how society will respond to more frequent and longer compound extremes over the next couple of decades.

Completeness of and Potential Geographic Biases in Current Understanding of Physical Tipping Points

Lenton expressed that current understanding of physical tipping points is likely below 50% and emphasized that there are many remaining known unknowns and unknown unknowns. He noted a likely clustering of understanding around the North Atlantic region, because it is a key part of the Earth system dynamics involved in past tipping points, leading to a potential bias toward this research topic and region. Lenton emphasized that physical understanding of climate tipping points is one of many important aspects of a well-rounded climate risk assessment. Other aspects have received less attention, such as potential impacts, vulnerability, and insights for risk management for society and decision-makers. Kopp added that current climate models suggest that Southern Ocean cloudiness may influence climate sensitivity and emphasized that it could be another potential area on which to focus tipping points research. Kopp noted that there are overlaps between what are considered tipping points and low-likelihood high impact events and outcomes, and how that body of work can also help characterize current understanding.

Modeling Tools and Quantitative Understanding

A participant inquired about the sufficiency of current modeling tools and quantitative understanding of processes related to tipping points and social impacts to support assessment of climate risks associated with transient high-overshoot scenarios. Lenton responded that current models are useful, but the room for improvement is substantial, especially regarding the potential social impacts. He suggested the usefulness of mapping out the potential physical impacts of tipping point potential scenarios to social impacts related via ecological and resource system nonlinearities and linearities. Kopp and Rubin also emphasized the lack of meaningful modeling for potential social impacts and the urgency of mapping out potential impacts, exposure, and cascading risks. Kopp highlighted the challenge of uncertainties in climate models and suggested a focus on adapting to using storyline approaches rather than narrowing down uncertainties. Rubin pointed out the challenges in making adaptation decisions due to a lack of meaningful standards beyond the built environment and suggested there is a need to determine useful baselines. Both Lenton and Kopp acknowledged the challenge of assessing risks associated with transient high-overshoot scenarios, given uncertainties in Earth system responses, and the need for a qualitative understanding of storylines.

Tipping Into Instability

Regarding the possibility for a tipping point to tip toward an unstable state, Lenton clarified that although, technically, the system will always find some stable equilibrium, reaching that equilibrium may take a very long time. Therefore, tipping a system may lead to a chain of events that continue for a long time before they settle, which on shorter timescales may appear unstable. Lenton discussed the possibility of the current climate state tipping into a different stable climate state and emphasized the importance of ongoing research in ruling out such scenarios.

The Q&A session highlighted the complexities and challenges associated with tipping points in both Earth systems and societal contexts. The speakers emphasized the importance of continued research to improve modeling, understanding potential social impacts, and addressing uncertainties regarding, for example, scenario plausibility and interconnectedness and coupling of systems and their potential impacts and tipping points. The session stressed the importance of relating tipping points research to societally relevant decision and actions with which decision-makers and policy-makers are currently grappling.

2
Historical Analysis of Past Biogeophysical and Social Tipping Points

Following the introductory remarks and overview of tipping points, the first topical session of the workshop highlighted examples of historical analysis of past biogeophysical and social tipping points from a range of perspectives and approaches. Workshop participants first gathered in plenary to hear presentations on these examples and then joined breakout group discussions to dive deeper into biogeophysical and social tipping points that occur over different timescales.

ECONOMIC PERSPECTIVE ON SOCIAL TIPPING POINTS RELATING TO CLIMATE CHANGE

Dr. Ilan Noy, Victoria University of Wellington, focused on the economic perspective of tipping points associated with climate change. He shared highlights from Newman and Noy (2023) that considered extreme weather events, such as heat waves, floods, and extreme cold, as significant impacts of climate change on the economy. He explained that the study analyzed around 250 attribution studies on extreme events around the world and matched the economic costs of these events, when available, to estimate the current cost of climate change. In doing so, Noy emphasized that the study revealed the increasing likelihood and economic costs of heatwaves and floods due to climate change. By matching the fractions of attributable risk to economic costs, Noy explained that the study also showed that over the past 20 years, global gross domestic production was affected by almost 1 percent, with impacts to the United States costing around US$150 billion per year. Therefore, Noy stressed that extreme weather events, often considered low-likelihood high-impact events, are already significantly impacting the global economy.

Next, Noy focused on the damages from extreme events, such as mortality or asset damage, that result in declines in economic activity with longer-lasting effects on the impacted communities. Noy identified this space as where tipping points start to matter. Noy provided two examples of potential scenarios following a disaster: recovery to pre-event conditions or a permanent decline in economic growth (Figure 2-1). He emphasized that tipping points play a role when an extreme event can lead to institutional changes that result in a different economic trajectory, which can be beneficial or catastrophic. Noy used a historical example of the 1978 earthquake in Iran, which triggered the Islamic Revolution and subsequent economic decline, to illustrate how a single disaster event can initiate a cascade of dynamics that leads to long-term economic consequences.

Historical Analysis of Past Biogeophysical and Social Tipping Points 15

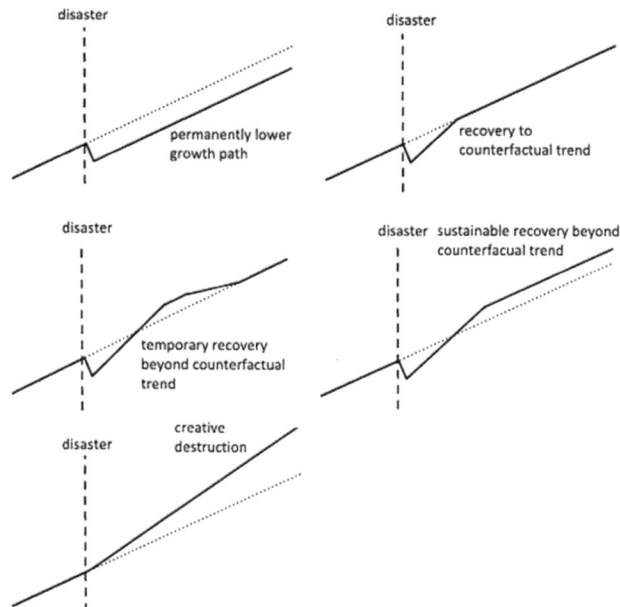

FIGURE 2-1 Potential scenarios of time evolution following a disaster.
SOURCE: Heger and Neumayer (2019).

Noy provided three more historical examples of past extreme events or disasters to demonstrate the complexity and broader impact of social tipping points triggered by these events and the wide range of resulting possible socioeconomic trajectories. Noy described the lasting economic and social impacts of the Dust Bowl in the United States; the beneficial, though not long-lived, political regime changes triggered by tropical cyclone Nargis in Myanmar; and the drought in the Levant, which may have led to the Syrian civil war and the migration crisis across Europe. Noy concluded by emphasizing a key point—following perturbations related to extreme events caused by climate change, social systems might not necessarily revert to a settled equilibrium, and therefore have the potential for prolonged and unpredictable impacts on economic and social systems.

HISTORICAL EXAMPLE OF A TIPPING POINT: THE DUST BOWL 1930S—THE PERFECT DISASTER

Dr. Benjamin Cook, National Aeronautics and Space Administration Goddard Institute for Space Studies, spoke in more detail about the Dust Bowl drought in the United States and its profound agricultural and societal impacts during the 1930s and beyond. Cook opined that the Dust Bowl was most likely the most severe disaster in U.S. history, causing widespread crop failures, land abandonment, and human migration. Cook also noted that although droughts and dust storms in the Central Plains were not uncommon before or after the Dust Bowl, the unique severity and scale of the impacts of the Dust Bowl were due to a combination of "bad luck and bad choices," with contributing factors ranging from the physical climate system, the regional ecology, and the choices people made.

Cook shared contributing factors in the century leading up to the Dust Bowl (1800s and 1900s). He noted that prior to the Civil War in the late 1800s, most agriculture in the United States was located east of the Mississippi River. Following the Civil War, the U.S. government undertook a campaign to remove and displace Native peoples across the Central United States, which opened the Central Plains for settlement and shifted agriculture from the eastern United States to the central plains. To cope with the relatively drier conditions of the Central Plains, agriculture moved toward crop types and practices that preserved water and drew up water from deep below the surface, relying on shallow-rooted, drought-sensitive crops such as wheat and corn. Cook discussed a shift toward the Campbell Method, where subsurface soil was tightly packed to draw up moisture from deep below the surface and covered by a looser surface soil mulch layer, covered by leaves and loose soil to reduce evaporation. With the advent of mechanization, the use of one-way disk plows became widespread to loosen up surface soil, which inadvertently contributed to soil erosion and left the Central Plains vulnerable to wind erosion. Shallower-rooted crops and looser soil surfaces replaced native, wild-adapted, deep-rooted crops of the Central Plains, further propelling the replacement of native resilient ecosystems with erosion- and drought-prone crops.

Cook explained that because these shifts coincidentally occurred during an extremely wet period, negative consequences were not immediate. However, a shift in ocean patterns, which typically drive droughts in western North America, triggered a cascade of events that led to prolonged and intense drought in the Central Plains. He noted that the resulting large-scale crop failures coincided with an economic depression, which led to widespread abandonment of farmland and further drying of the lands and providing the conditions to generate devastating dust storms.

Cook presented a discussion of feedbacks between dust and land degradation in agriculture. He explained that during drought, as vegetation is being removed from the land surface, evaporation of moisture into the atmosphere is reduced. Devegetated dry surfaces allow for loose surface dirt to be picked up by the wind, where the dust aerosols stabilize the atmosphere and suppress precipitation. This process leads to further loss of vegetation and to further surface drying.

Cook then shifted focus to the end of the Dust Bowl, including reevaluation and increased involvement of various government agencies to prevent reoccurrences. Cook shared two important efforts to mitigate the Dust Bowl's effects: soil conservation measures and increased irrigation. These interventional methods helped prevent similar disasters in the future and increased drought resilience in the Central Plains.

Cook emphasized that the Dust Bowl represented a tipping point resulting from a unique combination of natural and human factors and decisions. He stressed that lessons learned from this event highlight the important role that improved management practices play in reducing the likelihood of similar events occurring in the future.

A SIMPLIFIED MODEL OF BIOGEOPHYSICAL TIPPING POINTS: LOVELOCK'S DAISYWORLD

Next, Dr. Tim Lenton, University of Exeter, presented prepared remarks of Dr. Lee Kump, Pennsylvania State University. Lenton noted that Kump's presentation would focus on the concept of biogeophysical tipping points, particularly on how tipping point dynamics may arise from certain biological responses to environmental variables. Lenton explained that Kump's remarks would review the work of Jim Lovelock and Lynn Margulis and their Gaia hypothesis (Lovelock, 1972; Margulis and Lovelock, 1973), which proposes that life and the nonliving parts of the planet are coupled together in such a way that the Earth's environment is self-regulating over geological timescales. Lenton elaborated that Lovelock interpreted the Earth's history as long intervals of stability interspersed with rapid tipping-point transitions. Lovelock noticed that biological systems tend to exhibit parabolic responses to environmental variables, as opposed to purely physical and chemical processes that tend to respond monotonically. Lenton noted that this response has important implications for feedback dynamics.

Lenton used Lovelock's Daisyworld model to elaborate on this concept. He explained that this model simulates a fictional planet and the interactions between its only two lifeforms—black daisies and white daisies. In Daisyworld, the color of the daisies affects the planet's temperature, with black daisies absorbing sunlight and warming the environment, while white daises reflect sunlight and cool the environment. He noted that the daisies' effect on temperature starts locally, but as the daisies spread far enough, the effect can scale up to affect the global environment. Lovelock modeled the daisies' growth response function to temperature parabolically to simulate an optimal temperature for growth. Lovelock seeded the planet with black and white daisies and let the model run forward with time, slowly increasing the solar luminosity to mimic the life cycle of a star and letting the daisy growth and temperature interact. Through such a simulation, Lovelock was able to simulate how the population dynamics of the two types of daisies in response to their coupled temperature responses can lead to tipping points, where he observed rapid system transitions between stable states (as demonstrated by the rapid transitions in areal coverage percentage of the black and white daisies in Figure 2-2a).

Lenton discussed Lovelock and Kump's joint work to extend the simplified Daisyworld concept to various real-world ecosystems on Earth, such as plants and algae, which also have parabolic responses to environmental variables such as temperature. Lenton pointed out that both plants and algae have the potential to play an important role in regulating the Earth's climate through their coupling to other environmental processes, such as carbon removal and cloud formation, which can lead to potential tipping points. Lenton explained that Lovelock noted that although these organisms have cooling effects for the Earth and contribute to climate stability, they can only do so up to a certain extent (Kump and Lovelock, 1995). If pushed past certain thresholds (e.g., if sea surface temperatures were to rise above 12°C), algae populations would collapse, and as a result, a tipping point would occur with the abrupt termination of their cooling effect on the climate and a rapid rise in temperature. Lenton also noted that although Lovelock's Daisyworld model allowed the daisies to spread globally, spatial heterogeneity in the real world may increase or dampen the risk of large-scale tipping dynamics, and therefore more research is needed to improve understanding of this topic.

Lenton concluded by highlighting the usefulness of simplified models to understand the complexity of and interconnectedness among physical, chemical, and biological systems in determining the stability, and therein tipping points, of the Earth's climate system. To this point, Lenton added that Kump concluded his prepared remarks by stating that because life both affects and can be affected by the environment, it is an integral part of the global-scale regulation of climate. Because of their parabolic response nature, the sign of the related feedbacks may change past certain thresholds, leading to tipping points.

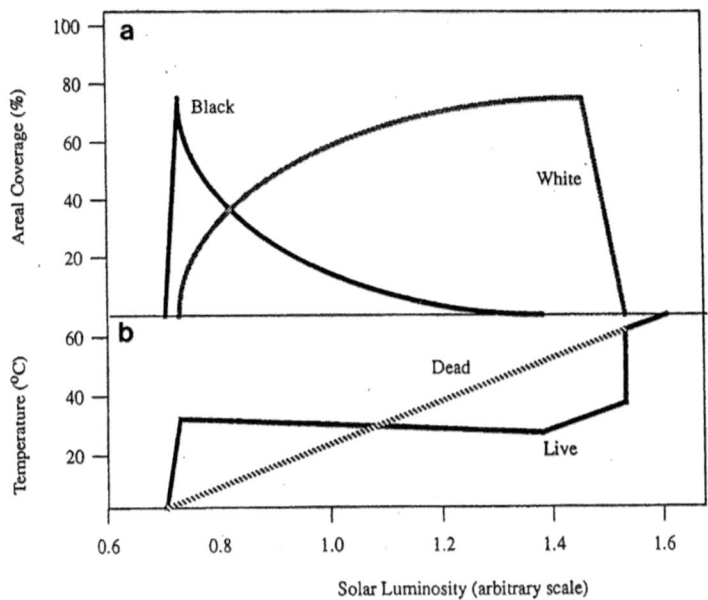

FIGURE 2-2 Evolution of global areal coverage percentage (a) and temperature (b) of black and white daisies in Daisyworld against solar luminosity.
SOURCE: Kump and Lovelock (1995).

BREAKOUT GROUP DISCUSSIONS ON BIOGEOPHYSICAL AND SOCIAL TIPPING POINTS

Following these presentations, workshop attendees participated in short breakout discussions. The participants were divided into three groups by timescale: (1) Larger-Longer Scale, (2) Cascading Risks Last 1,000 Years, and (3) Last 100 Years. Each group was asked two questions pertaining to their timescale focus:

1. What are the key outstanding research questions on biogeophysical and social tipping points, the interacting risks of these tipping points, and their cascading impacts?
2. What are the major barriers and opportunities to accelerate progress to advance these areas of research?

Following each breakout session, designated rapporteurs and participants from each group shared the groups' key takeaways.

Larger-Longer Scale

Dr. Simon Dietz, London School of Economics and Political Science, shared his group's key takeaways from its discussion on the larger-longer scale risks of tipping points. Dietz emphasized that the larger-deeper timescale perspective of tipping points is important for understanding the stability of the Earth system when subjected to various forcings. He noted that many open key research questions regarding the stability of the coupled natural and physical systems remain. Dietz shared that the breakout discussion focused on the social aspect of tipping points and the importance of gaining a better understanding of the pertinence of the deep-time perspective for social systems and social science questions. He closed his remarks with one opportunity and one barrier regarding the larger-longer scale. The opportunity was the potential that higher-resolution data may offer to improve modeling and understanding, while lack of incentives for transdisciplinary research concerning such long timescales relative to those of other social science disciplines was a barrier. Lenton added that another recurring theme of the group's discussion centered on how social dynamics may cause tipping in natural systems and vice versa. He highlighted that the group was drawn toward improving understanding of the longer-term perspective of humans in the Earth system and clarifying the importance of new social theories to address fundamental questions around modern growth regimes and the role of societal dynamics.

Cascading Risks Last 1,000 Years

Dr. Michael Schoon, Arizona State University, shared this group's key takeaways about tipping points in the most recent millennium. On understanding social tipping points, Schoon emphasized that his group discussed that these are not well understood, partly due to lack of fundamental understanding and partly from the challenge of reconciling various timescales across biogeophysical records from millennia ago to human's day-to-day lives and generational perspectives. He mentioned that the group discussed how further work may be needed to study the intersection of social and biogeophysical tipping points and that tools such as integrated assessment models may need to reexamine how well social processes are represented. Schoon pointed out that a barrier to this work is the discrepancy between how well-defined tipping points in biogeophysical systems are compared to those in social contexts.

Dr. Kristen St. John, James Madison University, added that the group discussed the usefulness of conceptual models to facilitate transdisciplinary team building and communication with stakeholders, particularly in helping to illustrate cause-and-impact relationships between human and natural systems. She also noted that the group discussed the importance of more research on phase changes in the physical system and their feedbacks onto both biogeochemical and human systems. St. John stressed the importance of understanding both the positive and negative impacts of tipping points and their role in providing information for decision-making that considers environmental, social, and climate justice issues.

Finally, Dr. Jeffrey Rubin offered two key takeaways: (1) the importance of considering political cycles when identifying barriers and opportunities for cascading risks research at the millennia scale and (2) the importance of clearly communicating the distinction between inevitability and imminence when considering short-term actions on a human scale using insights from millennia-scale tipping points.

Cascading Risks Last 100 Years

Margo Corum, staff member with the National Academies of Sciences, Engineering, and Medicine, summarized the takeaways of the group discussion on tipping points and risk within the most recent 100 years. She shared that this group elevated the importance of understanding societal amplification dynamics, which may allow natural disasters to scale up to human disasters. Corum noted opportunities for observing this dynamic in the evolution of societal changes from the Middle Ages to the Renaissance. Corum shared the group's discussion on understanding how societal systems respond to and enforce rates of change to inform policy decisions and risk assessment. Regarding opportunities, Corum highlighted the potential for new technologies and tools, such as improved sensor technology, modeling, and risk forecasting, to improve communication. Corum added that the group's discussion on risk management also involved contrasting historical examples of unmanaged abandonment, such as the Dust Bowl, to the modern concept of managed retreat. This discussion, Corum noted, highlighted the importance of planning and proactively strategizing to mitigate risks and adapt to environmental change.

Altogether, key recurring themes across the breakout groups' discussions centered on the complexity of cascading risks and tipping points across disciplines and timescales, the importance of improved communication and transdisciplinary collaborations, and the number of remaining fundamental questions that may involve integrated approaches to account for the complexity of the natural and human systems.

3

Regional Perspectives on Climate Tipping Points and Cascading Risk

The second day of the workshop focused on climate tipping points and cascading impacts and their interacting risks, highlighting different regional perspectives. The workshop participants first gathered in plenary to hear presentations about different geographical regions, including the coastal United States, American West, Arctic, and the Great Plains. The second part of the day included opportunities for broader participant interaction with a set of breakout sessions, followed by discussion on key takeaways.

COASTAL UNITED STATES: TIPPING POINTS IN FUTURE TROPICAL PACIFIC ISLAND SUSTAINABILITY

The day's first speaker, Dr. Curt Storlazzi, U.S. Geological Survey (USGS) focused on the impacts of climate change on Pacific atolls, specifically sea level rise and wave-driven flooding. Atolls are low-lying, ring-shaped features built up on coral reefs, and Storlazzi explained that they have significant strategic, economic, and ecological importance because of their critical infrastructure, biodiversity, and locations within the nation's Exclusive Economic Zones. Some examples include the northwestern Hawaiian Islands, the Marshall Islands, the Phoenix Islands, and the Gilbert Islands.

Storlazzi stated that unique characteristics of atolls leave them vulnerable. Atolls have very low topography with elevations of only 1 to 2 meters and steep offshore bathymetry. Each island has a limited freshwater lens, which, Storlazzi explained, provides the main source of freshwater and is critical for human habitability and the survival of many native endangered species. Infrastructure, critical habitats, and population centers of atolls lie at extremely low elevations. He explained that these characteristics make studying their risks and vulnerabilities translatable to other islands, such as American Samoa, Guam, Puerto Rico, and the U.S. Virgin Islands, because main population centers, infrastructure, and critical habitats are at low elevations along their shorelines.

Storlazzi next highlighted two current concerns related to atolls and low-elevation shorelines of higher islands. He noted that historically, wave-driven overwash events would occur every 20 to 30 years but are now happening multiple times per decade. This increase is attributed to rising sea levels, which have been increasing two to three times faster in the Western Pacific relative to the global average (Figure 3-1). Although a portion of this sea level rise is also driven by natural modes of variability, such as the Pacific decadal oscillation, Storlazzi noted with sea level rise of over one foot since 1990, implications for atolls that lie only 3 to 4 feet above sea level will become more severe.

21

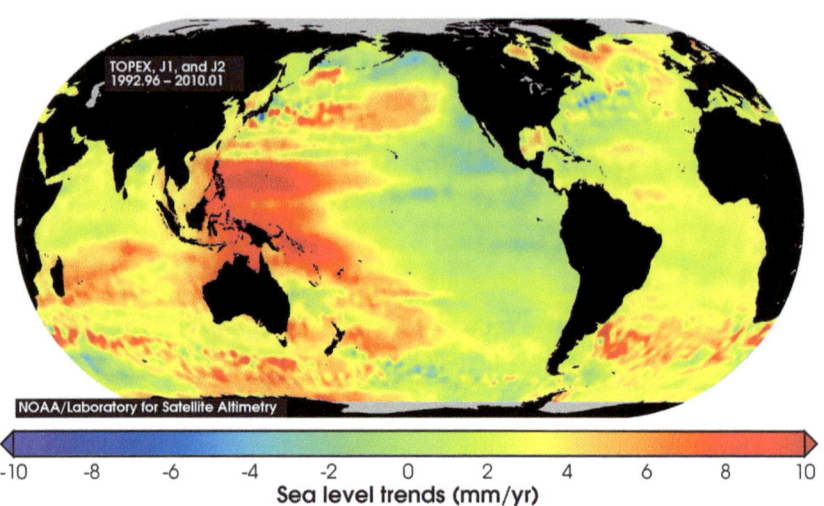

FIGURE 3-1 Sea level trends (mm/year) averaged over 1992 to 2010 from the NOAA Topography Experiment (TOPEX).
SOURCES: Presented by Curt Storlazzi on January 18, 2023, using NOAA sea level rise data from TOPEX (Fu et al., 1994).

The second concern Storlazzi emphasized is the interaction of sea level rise and waves. He pointed out that sea level rise through 2100 is expected to be 2 to 10 times higher than vertical reef growth rate (Montaggioni, 2005), which will lead to amplified wave-driven flooding and nonlinear flooding impacts, illustrated in Figure 3-2. This will cause greater inundation of infrastructure, freshwater supplies, agriculture, and habitats, Storlazzi emphasized.

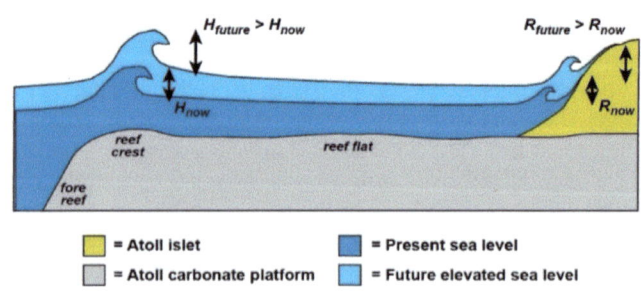

FIGURE 3-2 Relationship between sea level rise, wave height, and shore run-up.
SOURCE: Storlazzi et al. (2018).

Storlazzi presented his team's work to assess the resulting impact of sea level rise and its nonlinear interactions with storm wave-induced flooding on island communities (Storlazzi et al., 2018). His team used a suite of models, including general circulation and regional climate models, a wave-driven flood model, and a freshwater availability model to project future wave-driven events under a range of sea level rise and climate change scenarios to determine tipping points, such as a loss of infrastructure viability and freshwater availability. They calibrated their models with field data, focusing on modeling typical storms with magnitudes that are expected to occur annually under future sea level rise scenarios. Storlazzi presented the key findings of Storlazzi et al. (2018), which identified that tipping points occur when wave-driven flooding events occur more frequently than freshwater lenses can recover, resulting in a loss of freshwater availability and, hence, loss of human habitability. Their results indicate such tipping points could occur as soon as 10 to 60 years from now. Storlazzi added that their simulations projected that in most cases, typhoons would become less frequent, resulting in less rainfall and smaller waves on annual scales. However, he pointed out that those typhoons are projected to be much more intense, which would suggest increased vulnerability for these islands.

Storlazzi emphasized that many atoll islands may become unsustainable for human habitation due to increased frequency and magnitude of wave-driven flooding from sea level rise. Many atoll islands may not be able to invest in expensive adaptation measures, such as seawalls, structure elevation, and desalination plants, he said. Without these adaptation measures, Storlazzi explained that the potential loss of habitability may force inhabitants to evacuate, which can potentially raise significant geopolitical questions around Exclusive Economic Zones and nationhood. Storlazzi ended by highlighting the urgent need for proactive adaptation strategies and international cooperation to address the vulnerabilities of atoll communities to climate change impacts.

AMERICAN WEST: SUMMER WATER IN THE WEST—CLIMATE TIPPING POINTS AND CASCADING IMPACTS

Next, Dr. Gordon Grant of the U.S. Department of Agriculture Forest Service and Oregon State University presented on challenges surrounding water management in the American West, with a focus on seasonal distribution of water, climate change impacts, and innovative solutions. Grant shared that his presentation was motivated by the paradox of water in the West, where precipitation primarily occurs during winter but agricultural, urban, and ecological water demand peaks during the summer (Figure 3-3). Grant indicated that the disconnect between the seasonality of water availability and demand requires effective storage solutions to bridge this gap. He outlined various storage mechanisms, both natural and engineered, that are critical for managing Western water resources, including snowpack, dams and reservoirs, and groundwater (Figure 3-4). He stressed that understanding the spatial and temporal dynamics of these storage systems is key to effectively manage water availability throughout the year, where their interplay may lead to tipping points and cascading impacts.

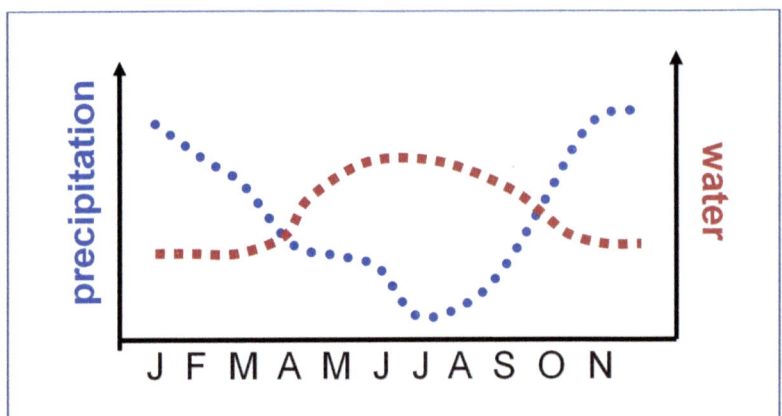

FIGURE 3-3 Paradox of water in the West, comparing precipitation (blue dotted line) and water demand (red dashed line).
SOURCE: Presented by Gordon Grant on January 18, 2023.

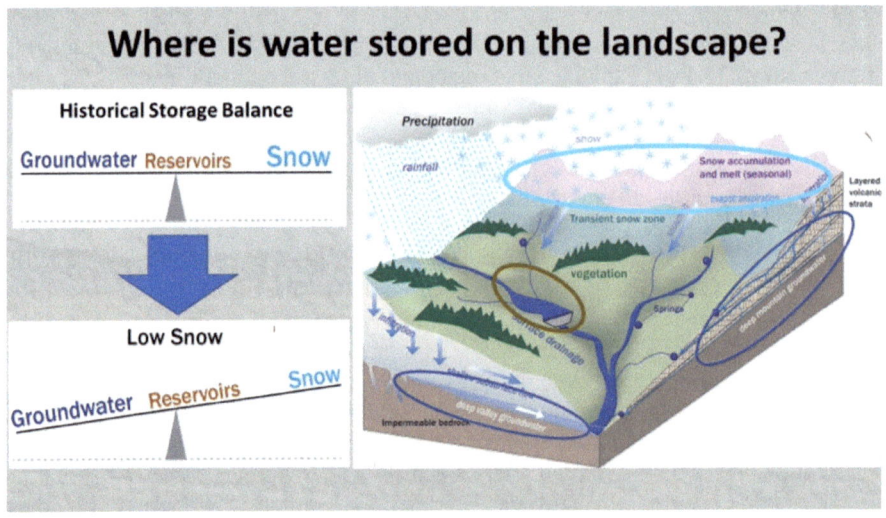

FIGURE 3-4 Natural and engineered water storage mechanisms for Western water resources.
SOURCE: Presented by Gordon Grant on January 18, 2023.

Grant emphasized that perhaps the most alarming trend and first potential tipping point in the future is diminishing snowpacks due to rising temperatures driven by climate change. He highlighted the rapid rate at which snowpacks are diminishing in both observations and models under median emission scenarios, with most models projecting the complete removal of persistent snow from watersheds by 2050–2060. He then stressed the complexity of predicting the consequences for water availability due to the different rates of diminishing snowpack across diverse topography and noted that further research is needed to understand snowpack interactions with rivers, oceans, and vegetation on water availability and storage. He encouraged an integrated approach, combining satellite data, fundamental understanding, and modeling approaches.

Grant then focused on the next impact of diminishing snowpack—reservoir storage. As snowpack diminishes, Western states are experiencing significantly diminished low flows for reservoirs, he said. Grant discussed the need for adaptive approaches to address the challenges, such as potentially using forecasts to inform reservoir operations or reconsidering the legal and institutional frameworks that govern reservoir management.

Grant raised concerns around the depletion of groundwater resources, stressing that this could likely lead to emerging tipping points. He explained that intense utilization of valley aquifers for agricultural use has dramatically depleted groundwater resources. He shared that recent groundbreaking techniques in remote sensing coupled with volumetric measurements have provided improved insights into groundwater availability during droughts but that there is still a need to improve understanding of rates of change and time trajectories to inform groundwaters management. Grant highlighted the importance of addressing extraction rates and exploring the efficacy of recharge strategies to sustain groundwater levels, especially during droughts. He noted that research in these areas may inform open questions and challenges regarding groundwater availability and implications for agriculture, contamination, and potential adaptation strategies.

Grant introduced mountain groundwater, where large volumes of water are stored within the bedrock in volcanic landscapes, as an underappreciated component of the Western water landscape. Volcanic aquifers act as reservoirs for winter precipitation and then make up the majority of late summer river flow, Grant explained. According to Grant, integrating volcanic aquifers into the water storage strategies may potentially provide a beneficial tipping point to help sustain regions that are facing challenges using solely traditional surface water storage strategies.

He indicated that the potential opportunity provided by mountain groundwater also comes with unknown challenges, for example, uncertainty about where, when, and how much mountain groundwater is available and potential interactions with landscape disturbances, such as vegetation and fire. He highlighted the need for interdisciplinary research and integrated approaches to address various aspects of the complex landscape of Western water management. Grant emphasized the importance of understanding water dynamics, developing forecasting methods, and examining the interactions between water systems and ecosystems. Grant finished by suggesting the establishment of regional centers dedicated to studying integrated water problems, where collaborative research efforts could coalesce around natural and social dimensions, cascading impacts, and innovative water management solutions to address the challenges posed by climate change and growing water demands in the American West.

ARCTIC PERSPECTIVES

Societal Tipping Points to Shape Sustainable Change

Dr. Terry Chapin, University of Alaska, presented on the social dimension of Arctic tipping points. He focused his remarks on triggering positive social tipping points that can lead to sustainable outcomes for both nature and society and identified human drivers, global population growth, and resource use as key factors contributing to environmental changes. He explained that these changes include warming temperatures and the conversion of natural ecosystems to managed ones, leading to consequences such as species extinction and overexploitation of resources. Chapin pointed out that changes in the environment and ecosystems affect the benefits that society derives from nature. Using Alaska as an example, Chapin highlighted the challenge in balancing economic interests like oil extraction (a significant source of the state's revenue), with growing climate change impacts on both nature and people in the region. He pondered how society might shift away from tipping points that further contribute to human drivers of environmental change toward tipping points that lead to a more sustainable trajectory.

Chapin introduced the concept of "stewardship" as a way for society to actively shape pathways of socio-ecological change, rather than respond to changes as they occur. He emphasized that stewardship involves intentional transformation that is oriented around goals for ecosystem health and human well-being. He asserted that these goals, while often posed as a choice between people and nature, are inextricably linked because humans are an integral part of nature and cannot, in the long term, benefit without interventions that support nature. Chapin posed two questions:

1. Can tipping points toward desirable changes be deliberately triggered?
2. If so, what are the desirable changes society would like to see?

He noted that deliberately triggering transformation may lead to both opportunities and risks and highlighted the importance of equity considerations.

Chapin outlined Olsson et al.'s three-phase framework for transformation: preparation, navigating change, and building resilience of the new system (Olsson et al., 2004). He explained the importance of defining the goals and strategic steps to determine the transformation and its trajectory during the first preparation phase, with scenario analysis being a helpful method. Chapin described the importance of identifying barriers, potential triggers for change, and potential allies for navigating change. To illustrate, Chapin provided the example of the war in Ukraine creating an opportunity for Europe to move away from dependence on Russian oil and toward development of renewable energy. Chapin added that being flexible in strategic approaches during the navigating change phase facilitates adaptation to the potential range of directions of change and maximizes the proportion of society that can be engaged in sustainability efforts, which increases the likelihood of the transformation success. Chapin added that building resilience into the new system in the third phase is also critical to prevent the new system from reverting to the old system.

Chapin drew from Donella Meadows's work to describe different levers for triggering socio-political tipping points (Meadows, 2015). He noted the tradeoffs between the depth and scope of the transformation and the ease of its implementation (Figure 3-5). Chapin indicated that there are multiple possible pathways to create socio-political tipping points that could involve interactions between different levers. As such, multiple strategies tailored to specific contexts are warranted, Chapin emphasized. He described potential strategies that include shifting norms and behaviors, incentivizing sustainable production and consumption, and empowering society for change.

On shifting norms and behaviors, Chapin emphasized the importance of broadening society's vision, one of the most difficult levers for triggering socio-political change. He described the societal focus on profit, material wealth, and comfort in the 19th and 20th centuries and stressed the need to shift societal values toward prioritization of human well-being and nature conservation. Chapin described how the stewardship framework may support this shift, because it focuses on developing complementary dimensions of human well-being, for example, natural capital and human capital. Chapin then discussed a set of survey results (Gaffney et al., 2021) from developed nations that suggest that most people believe human actions are driving toward a tipping point at the global scale and now favor a shift in focus from societal goals built around human well-being rather than profit. Despite this desire, the challenge of translating these values into actions remains, Chapin added. Chapin highlighted the important role of education and identity formation in shifting norms. He noted that indigenous groups in Alaska, though small in proportion, are widespread, and often have important values oriented around protecting nature. Showcasing their sustainable behaviors can increase awareness of key identities and values and encourage broader societal movement toward sustainability, he said.

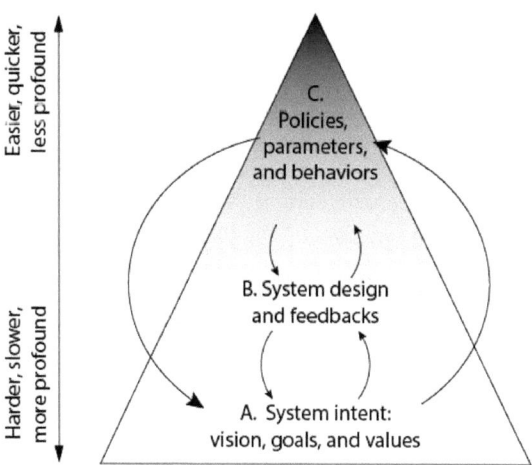

FIGURE 3-5 Tradeoffs between depth and scope of transformation and difficulty in implementation.
SOURCE: Chapin et al. (2022).

Chapin described approaches to incentivizing sustainable production and consumption that could involve making structural adjustments, such as changing costs and benefits through incentives and disincentives, or taxes and subsidies, and improving rules around sustainable actions. In addition, he suggested that society can be empowered to create positive change toward sustainability by engaging nontraditional influential actors, fostering corporate citizenship, and exploring new democratic institutions. Chapin provided examples of this approach, including citizenship for nature as opposed to corporations, and citizen assemblies as opposed to conferences of parties for international negotiations. Finally, Chapin stressed the importance of initiating change at the local level, starting with individual behaviors, community engagement, and emphasizing co-benefits to garner larger public support and foster sustainable transformation.

Impacts of Arctic and Boreal Tipping Elements at Local and Global Scales

Dr. Michelle Mack, Northern Arizona University, focused her remarks on Arctic and boreal tipping points, with an emphasis on terrestrial elements and their impacts on climate through the carbon cycle. Mack highlighted that elements such as sea ice collapse and thawing permafrost are amplifying human-driven climate change and that the rate of warming is propelling elements toward tipping points with the potential for cascading risks through self-perpetuating positive feedbacks. She emphasized the complexity and interconnectedness of these elements within the Earth system and their potential to drive fundamental changes in global climate.

Mack detailed her focus on two elements of terrestrial ecology: permafrost (soils and sediments frozen for more than 2 years up to millennia) and boreal forests. She emphasized how both elements play critical roles in the carbon cycle because they hold globally significant reservoirs of organic carbon, on the order of petagrams (Pg; 10^{15} g) (Figure 3-6). Mack explained that permafrost and boreal forests exchange carbon with the atmosphere through photosynthesis and respiration, but these exchanges occur with substantial time lags. She noted that as the soil warms, carbon that was stored back in the Pleistocene (2.58 million to 11,700 years ago) may be released today. Mack stressed that the massive size of the carbon reservoirs represented by permafrost and boreal forests indicates that even small increases in the rate of exchange between these systems and the atmosphere can have globally important impacts on atmospheric carbon. Mack underscored the sensitivity of these carbon pools to climate warming, explaining that these exchanges are determined by "threshold behaviors." In the case of ice and permafrost, Mack noted the threshold between frozen and thawed states, and for forests, the threshold between old and new growth. She explained that permafrost thaw has the potential to range from a regional to a global tipping point because of the vast amounts of organic carbon stored. She contrasted this with boreal forests, which she suggested are more likely to contribute to regional tipping points. Mack noted the potential for these regional tipping points to have cascading interactions with permafrost due to unique system-level behaviors, which can lead to subsequent global impacts.

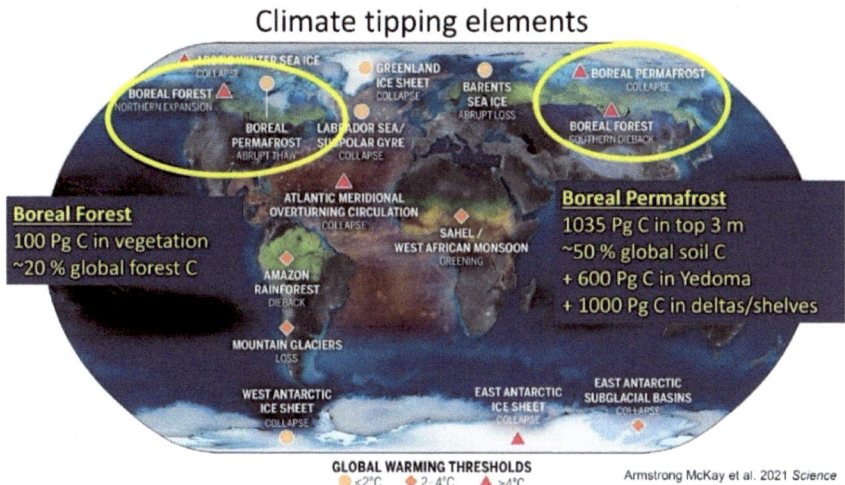

FIGURE 3-6 Climate tipping point elements associated with the boreal forests and permafrost and estimates of amounts of carbon (in Pg C) contained in each element.
SOURCES: Presented by Michelle Mack on January 18, 2023. Adapted from Armstrong McKay et al. (2022).

Mack described in more detail the threshold dynamics of boreal forest carbon cycling and its cascading impacts on permafrost. She discussed the complexity of the role of wildfires in boreal forests, which historically have been part of the natural disturbance regime but are increasing in frequency and intensity in response to the warming climate. These changes contribute to increased amounts of energy released within and the impacts on the affected ecosystems. Mack noted that as the climate changes in the Arctic, the potential for feedbacks between the biosphere and the atmosphere also increases. She explained that warmer and drier climate leads to increased lightning activity, triggering more fires. These fires transfer carbon from the ecosystems to the atmosphere, triggering additional warming in a self-perpetuating positive feedback loop. She stressed that fire affects both carbon inputs and outputs, and hence can lead to shifts in ecosystem carbon balance over the fire cycle (100 to 1,000 years). Mack emphasized the highly complex nature of the fire cycle, explaining that fires could also lead to effects that potentially reinforce or counter positive feedbacks through processes that impact albedo, evapotranspiration, vegetative composition, and fire-vegetation interactions. These feedbacks could either increase or decrease the intensity of fire on the landscape.

Mack explained that boreal coniferous forests serve as long-term carbon sinks. She noted that recent increasingly severe wildfires exacerbated by climate change have been burning deeper into the soil layer, reaching deeper levels of "legacy" carbon from pre-disturbed ecosystems that have built up over millennia. If this legacy carbon is unlocked, boreal forest ecosystems may rapidly switch from carbon sinks to carbon sources for the atmosphere, Mack explained. She described two other major impacts of this unlocking of legacy carbon on the long-term recovery of the carbon cycle: first, that historically dominant conifers are replaced by new species, and second, that permafrost can be exposed to higher air temperatures that accelerate its thaw. Mack compared two possible outcomes following severe fires that reach legacy carbon. She first described changes observed in Alaska, where fires reaching legacy carbon resulted in deciduous species replacing conifer, which can increase both carbon storage and resilience to fire. She contrasted the transitions observed in Alaska to those observed on the East Coast of the United States, where following a severe fire, the system failed to transition to deciduous species and coniferous species grew back. In these cases, the coniferous regrowth held less carbon but increased flammability. These two examples, Mack noted, highlight the range and complexity of the carbon and fire cycles in the boreal forest system.

Returning to permafrost, Mack explained that it is a legacy of previous climate and is not in equilibrium with the current climate because of the insulating effects from the organic layer of peat in the soil. She stated that in addition to gradual, top-down thawing, there is equal potential for rapid thaw. In these abrupt thawing scenarios, Mack explained that flowing water transports heat that creates concave areas in the ice, which traps warmer water against ice and leads to more ice melt—a positive feedback. Mack noted that within the next 100 years, wildfire is the most likely trigger for these abrupt thaws.

Mack presented different scenarios of future emissions of carbon dioxide and methane from permafrost carbon, illustrating how the amount and composition of these emissions highly depend on human-induced warming trajectories and their triggering and interactions with the complex feedback mechanisms of the boreal forest and permafrost elements. Mack then turned her remarks to what ecologists can do to impact these issues. From her perspective, she emphasized the need to reduce emissions and mitigate climate change to avoid the most severe impacts of permafrost thaw, the importance of broadening society's vision to address the challenges presented by climate change in the Arctic, and the role of global networks of scientists, such as the Permafrost Carbon Network, in communicating and educating the broader public and policymakers on the importance of Arctic processes. She also stressed the need for scientific diplomacy among the United States and other nations, because 70 percent of the terrestrial Arctic lies within Russia.

Mack ended by emphasizing the need for interdisciplinary approaches to address the biophysical dimension of Arctic tipping points. She highlighted the importance of integrating social perspectives with ecological understanding to develop effective strategies for mitigating climate change impacts in the Arctic region. She encouraged creative harnessing of knowledge of natural and human systems to design new systems that can enhance the ability to deliberately trigger positive, rather than negative, outcomes.

GREAT PLAINS: TIPPING POINTS AND CASCADING IMPACTS IN NEBRASKA

Dr. Craig Allen, University of Nebraska–Lincoln, focused his remarks on tipping points and cascading impacts in the American Great Plains. He highlighted the Sandhills and Platte River ecosystems in Nebraska. Allen noted that he would cover one example of a tipping point caused by a spatially contagious process[6] and another caused by surpassing threshold behavior.

The American Great Plains are a grassland biome threatened by several sources, Allen explained, including land use change for human settlements and agriculture and the alteration of key processes such as fire for maintaining grasslands.

Allen first focused on tree invasion, a global phenomenon occurring in grassland across the globe, including Mongolia and Australia, as well as in the Great Plains. The Sandhills are around 20,000 square miles of grass-covered dunes that comprise much of the land area in Nebraska. He detailed historical conditions in the Sandhills as recounted by Parker, an explorer in the region in the 1850s, who described the region as being dominated by bare ground covered in sparse grass. By the 1860s, the landscape dramatically changed. Natural bison were replaced by cattle, and roads and railroads created ignitions that triggered severe wildfires that cleared grass from the dunes, Allen explained. Allen pointed out how early settlers recognized that mobile dunes were a potential alternative stable state for the region when grass was removed. To manage the land to avoid this potential tipping point toward a less-desired sand dune state, management efforts were made to inhibit fires and reduce herbivore pressure from grazing, he said (Figure 3-7). Allen explained that currently, the land is so highly managed to reduce the possibility of mobile dunes that some herbaceous plants are now endangered. This heavy management occurs despite the vast size of the area and extremely low population.

Allen explained that windbreaks (rows of vegetation or physical barriers used to provide shelter from wind) are extensively employed in the Sandhills because dry sandy soil can endanger the health and livelihoods of people and livestock, a primary industry in the region. These windbreaks are predominantly comprised of Eastern Red Cedar, which Allen noted is one of the only tree species suitable for the region but is also extremely susceptible to fire. Allen described how red cedar has spread very quickly across the Sandhills region because of wind and animal seed propagation. Although the windbreaks provide an essential ecosystem service, Allen explained, Nebraska is currently losing around 40,000 acres a year to emerging tree-dominated ecosystems. These ecosystems replace grasslands, reducing grassland and livestock productivity. Allen stressed that removing the cedar is not sufficient to restore the grasslands because of the associated loss of seedbanks and soil properties necessary for grasslands to thrive. He added that red cedar systems may also release carbon more easily than grasslands because their susceptibility to and intensity of wildfires is higher. Current efforts are underway to reintroduce fire to the region to help restore grassland properties in areas invaded by the red cedar, Allen noted.

Allen then spoke about flow regime changes in the Platte River, which is charged by snowmelt and has historically experienced heavy spring surge flooding, making it both wide and shallow. These flow regime changes maintained unvegetated landscapes critical

---

[6] Spatial process that spreads contagiously, such as climate-sensitive disturbances like fire, land-use change, pests, and pathogens, and can drive changes in ecosystems, biogeochemical cycles, and land-atmosphere feedbacks (McCabe and Dietze, 2019).

Change in Grasslands

Ecological resilience: the capacity to withstand disturbance without shifting to an alternative regime.

Tipping Points and Surprise

Change and Coercion in Rivers

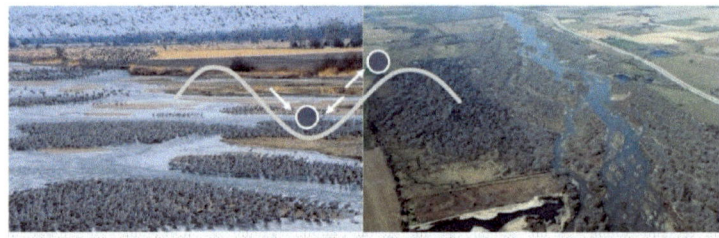

FIGURE 3-7 Transformation of Sandhills grasslands in Nebraska to an alternative regime following human management.
SOURCES: Presented by Craig R. Allen on January 18, 2023. Adapted from imagery from Chirs Helzer and the Nature Conservancy.

for natural wildlife species but also historically disrupted local agriculture, he noted. Human interventions such as dams were put in place to manage variability in the water height and provide opportunities for energy production for electrification and irrigation for agriculture, Allen added. He explained that these interventions reduced flow variability and led to an alternative stable state. Allen explained that removal of the spring surge flooding, the primary driving process of the river system, forced the system past a tipping point and toward steeper banks that eventually became vegetated, first with native species, but later with invasives and trees. The shift to wooded banks, Allen argued, tipped the Platte River into an alternative stable state that was largely beneficial to the human communities and agriculture. Ecologically, he pointed out, this new regime negatively impacted wildlife by removing natural environments of native bird species. Allen noted that in one actively managed 50-mile stretch of the river, efforts have been made to coerce the environment back to its pre-tipped state by cutting down vegetation and plowing sandbars. Although these actions help protect endangered species, Allen mentioned that these efforts are extremely costly ($200 million to date) and have been met with only partial success. Allen said that because of this program, the river exists in two states simultaneously, one a stable state created by previous interventions beneficial for humans and agriculture and another actively managed into a coerced state to benefit the native ecosystem. Allen stressed that many rare habitats are maintained entirely by this kind of coercive management, but these efforts remain costly and require continual maintenance to prevent the system from falling back into the surrounding stable state.

Allen then discussed how changes in grasslands are coupled with other global environmental regions and issues; for example, how reductions in productivity for grazing cattle in the Sandhills contribute to deforestation in the Amazon as beef production moves to those areas, or how changes to the Platte River increase agricultural productivity that contributes to hypoxic dead zones in the Gulf of Mexico. Allen emphasized that alterations in the grasslands, such as changes in groundwater extraction, nitrogen inputs, demographic characteristics of ranchers, and even market dynamics, all have the potential to be tipping points. Tipping these points, he added, may lead to cascading effects across the ecological, social, and economic systems that span local to global scales. Allen stressed that identifying tipping points is important, but anticipating the correct resulting stable state is challenging. He ended by reiterating that coercion can maintain a state when ecological systems are no longer sustaining those states on their own, but this strategy is expensive, temporary, and difficult to manage.

Q&A DISCUSSION WITH REGIONAL PERSPECTIVES: OVERARCHING THEMES

A moderated question-and-answer discussion followed each presentation in this session. The sections below summarize some of the key overarching themes from this discussion.

Integrated Approach

Across all of the discussions, speakers mentioned multifaceted and integrated approaches to address complex environmental challenges. They emphasized that scientific evidence, social dynamics, and ecological processes may be beneficial in decision-making and action. Storlazzi discussed advances in understanding the risks posed by melting glaciers and the need to connect these research efforts to adaptation efforts in the Pacific atoll islands. Concerning challenges with water availability in the West, Grant underscored the importance of interdisciplinary research and innovative strategies to carefully address water-related challenges in a sustainable and long-term way. Chapin and Mack highlighted the need for evidence-based decision-making when addressing environmental issues in the Arctic. They all highlighted the importance of research approaches that integrate observations, experiments, and scientific evidence and connect the findings to environmental management efforts. Allen discussed the complexities of ecological management in the Great Plains, the benefits of an integrated approach to decision-making around management, and the need to consider ecological processes, social dynamics, and ecosystem resilience.

Adaptation, Mitigation, and Resilience

All the speakers cited adaptation, mitigation, and resilience urged caution about as critical concepts in addressing environmental issues. Ecosystem resilience and social resilience were emphasized, as well as the importance of the ability for communities to adapt to changing conditions. Storlazzi noted that both mitigation and adaptation strategies are crucial to address the impacts of sea level rise in the Pacific islands. He emphasized the need to reduce emissions while also recognizing the urgency needed to implement adaptation measures for vulnerable island nations. Storlazzi also discussed the adaptation strategies being implemented in atoll communities, such as small-scale measures to build more resilient infrastructure and seawalls and large-scale efforts like investments to raise islands' elevation. Grant discussed the potential of mountain aquifers for adaptation strategies for resilience and highlighted the need to connect various adaptation strategies, such as forest and road management techniques, to capture snowmelt. Regarding building resilience in the Arctic, Chapin and Mack urged caution about the potential for unintended consequences of interference from adaptation and management efforts. Allen addressed the concept of ecosystem resilience, arguing that maintaining alternative stable states and thresholds can ensure resilience, and highlighted the complexities of balancing climate mitigation strategies with ecological considerations. He raised the potential for some strategies, such as afforestation, to have unintended long-term consequences, such as reducing resilience by negatively impacting local ecosystems and economies without having the desired long-term impacts on global carbon sources.

Communication and Engagement

Throughout the discussions, the speakers emphasized effective communication and engagement as essential components of efforts to address environmental challenges. Storlazzi noted that effective communication and collaboration will enhance sharing of research advances, as well as development and implementation of adaptation strategies relating to sea level rise and wave-driven flooding. He highlighted efforts to ensure that scientific research informs decision-making and action on the ground through engagement with local communities, government agencies, and international organizations. When asked by a participant about approaches to addressing misinformation, Chapin underscored the significance of the scientific community engaging with the public to provide compelling information that directly relates to solutions and potential benefits of moving toward sustainability. Along the theme of communication, the session's moderator, Dr. Dorothy Merritts, Franklin and Marshall College, highlighted the importance of effective science messaging and the use of clear and compelling storytelling to combat misinformation.

Uncertainty and Complexity

Each speaker acknowledged the uncertainties and complexities inherent in the environmental issues of their regions, particularly around future climate projections and the interconnectedness of ecological systems. Concerning the Pacific atolls, Storlazzi highlighted the uncertainty inherent in future scenarios regarding sea level rise and subsequent wave-driven flooding and their impacts. He pointed to the potential collapse of ice sheets as a major source of uncertainty when considering future scenarios and refining predictions to inform adaptation strategies in the Pacific atoll islands. Grant reiterated the complex and interconnected nature of the water system in the West. He acknowledged that the field is in the early stages of understanding of key processes such as groundwater recharge, spatiotemporal changes in mountain aquifer storage, and interactions between mountain aquifers and coastal water supplies, and stressed the need for advanced monitoring techniques to better address future challenges. Allen highlighted the interplay between social dynamics and ecological management strategies in the Great Plains, such as the complexity and uncertainty revolving around environmental management. He noted that drawing from their experience with the red cedar invasion, Nebraskans are hesitant to implement strategies such as afforestation. The speakers underscored the importance of ongoing research and monitoring to inform adaptive management strategies.

Balancing Human and Ecological Needs

A recurring theme throughout the speakers' discussions revolved around the importance of balancing human needs with ecological conservation goals. Speakers with various regional perspectives touched upon this difficult but necessary balance and emphasized the importance of reconciling competing interests, considering trade-offs, and finding collaborative solutions that benefit both people and the environment. Storlazzi underscored the necessity to consider cost-benefit ratios for adaptation efforts in the Pacific atoll islands, which typically have smaller economies, as well as the importance of considering environmental justice issues. Grant noted that dam removal has both ecological- and energy-related considerations and discussed the need to reconsider the role of reservoirs in water management as environmental and energy challenges evolve. Allen described federally funded restoration and governance initiatives aimed at restoring the Platte River back to its natural pre-managed state, which also brings ecotourism to benefit local communities ecologically and economically. Allen suggested that such collaborative approaches to river management could consider both human and ecological needs.

The speakers' discussions reflected a nuanced understanding of environmental challenges relating to tipping points and cascading impacts and recognition of the complexity and interconnectedness of social, ecological, and economic systems in addressing them effectively.

BREAKOUT GROUP DISCUSSIONS ON REGIONAL PERSPECTIVES ON CLIMATE TIPPING POINTS AND CASCADING IMPACTS

Workshop participants then had the opportunity to select one of four breakout discussions that focused on climate tipping points across each of the different regions presented earlier in the day. A rapporteur from each group was assigned to provide a brief summary of the key themes from their group's discussion.

Arctic Breakout Discussion Themes

Dr. Tim Lenton, University of Exeter, summarized the discussion on tipping points in the Arctic. One theme he highlighted was the need to look across both natural and social systems in holistic ways to identify potential tipping points and associated cascades. Lenton noted that the group discussed the importance of "redrawing a map and a list of potential tipping points," emphasizing that these tipping points would extend beyond physical systems into social ones, particularly for indigenous communities. Lenton also discussed coproduction of Western science and indigenous knowledge systems to better manage risks and seek positive social tipping points.

Coastal United States Breakout Discussion Themes

Dr. Kristen St. John, James Madison University, summarized the discussion on tipping points in the coastal United States. The first theme she focused on was on the importance of communicating science to ensure that the results are meaningful, and that stakeholders and community members are involved in the process from the beginning, which aligns with an identified need for regional coproduction of knowledge shared by Lenton. St. John also described the group discussion on uncertainty and unknowns and how to represent them in risk communication. She highlighted that something that is high risk but has many associated unknowns is not necessarily a low-likelihood event, which could introduce difficulties for decision makers. St. John noted that the breakout group also discussed the theme of transdisciplinary nature-based solutions for coastal areas that would have social and ecological co-benefits.

American West Breakout Discussion Themes

Dr. Michael Schoon, Arizona State University, summarized the breakout discussion on tipping points in the American West. Relating his comments to the previous discussion of dramatic feedbacks in the Arctic, Schoon described a discussion around cascading impacts of drought through a cycle of water stress, fire, and additional emissions that feed back into more severe drought conditions. He mentioned a similar process for extreme heat and surface water temperatures. Schoon also highlighted the group's discussion around transdisciplinary research opportunities, including the importance of a regional focus on infrastructure investments, geohealth, and the food-energy-water nexus.

Great Plains Breakout Discussion Themes

Dr. Simon Dietz, London School of Economics and Political Science, summarized the discussion on tipping points in the Great Plains. He highlighted several themes that were presented by the groups, including tipping points identified by Craig Allen in his presentation of the Sandhills and the Platte River, groundwater in the Ogallala aquifer, nitrogen thresholds for river systems, and precipitation thresholds for rain-dependent agriculture. Dietz noted that the group discussed research that has the capacity to examine teleconnections[7] among environmental and socioeconomic factors between the Great Plains region and other U.S. regions and beyond.

Online Breakout Discussion Themes

Jonathon Tucker, staff member with the National Academies of Sciences, Engineering, and Medicine, summarized the discussion in the breakout group for online attendees. He highlighted challenges in balancing economies and land use in the Great Plains and coupling remote sensing capabilities with a variety of different model types (e.g., downscaled global climate models, regional precipitation models, groundwater models) to identify the potential for water storage in vadose zones, volcanic zones, and depleted aquifers.

Key Cross-Cutting Themes

Although each breakout session focused on a particular region, some key themes discussed in the plenary included:[8]

1. ***Holistic Approaches*** that consider both natural and social systems
2. ***Community Engagement and Coproduction*** from the outset
3. ***Uncertainty Management*** to improve risk communication for decision makers

---

[7] A teleconnection broadly refers to a cause-and-effect relationship between remote regions caused by meteorological, societal, and/or ecological phenomenon. In physical Earth science, teleconnections refer to climate links between geographically separated regions (Nigam, 2003). Societal teleconnections are analogous to physical teleconnections, but rather refer to human-created linkages that connect activities, trends, and disruptions across different geographically separated social systems, communities, or regions (Moser and Hart, 2015).
[8] This summary of themes from the breakout session reflects the discussion of the group and should not be construed as reflecting consensus of the group.

4. ***Transdisciplinary Opportunities*** to explore social and ecological co-benefits
5. ***Identifying Regional Tipping Points*** to inform effective adaptation strategies
6. ***Research Priorities*** that leverage observation and modeling capabilities

4
Examples of Interdisciplinary Research Priorities and Opportunities

The final day of the workshop focused on discussing examples of interdisciplinary research priorities and opportunities. The morning session began with two speakers presenting on Indigenous Knowledge (IK) and environmental justice perspectives. The afternoon consisted of interactive breakout sessions that provided the opportunity for participants to brainstorm advancements in tipping points, cascading impacts, and interacting risk.

INSIGHTS FROM INDIGENOUS KNOWLEDGE

Dr. Kyle Whyte, University of Michigan, provided an overview of insights from IK and why it provides a unique perspective on how to consider and address climate tipping points. Whyte, an enrolled member of the Citizen Potawatomi Nation, works extensively with various organizations and initiatives that focus on environmental and climate justice. Whyte defined IK as empirical scientific traditions of different indigenous peoples, which differ from community to community but interact with one another in global dialogues. He acknowledged that multiple definitions of IK exist. IK is a social science, Whyte explained, because of the intersection of humans and ecosystems as part of the cluster of relationships that comprise the Earth system. Whyte described how historically IK has been ignored or even suppressed, but efforts are under way to uplift indigenous scientists and knowledge keepers. Although IK is often considered local knowledge, Whyte described it instead as knowledge that has been denied the opportunity to be in dialogue with people who have the privilege to do science that is global in scope. Whyte emphasized that efforts to advance the role of IK aim not only to use local IK to supplement other existing work, but also to bring IK holders to the table.

Whyte noted that IK can be thought of as a framework that coordinates knowledge about natural processes with approaches to managing and interacting with those processes. He highlighted that a central concept of IK is that a well-coordinated society that can respond to an emergency or a crisis is the product of years or even generations of trust building. Whyte explained that kinship relationships based on high degrees of trust, consent, accountability, reciprocity, and responsibility are critical because they enable people to coordinate and work together, especially in emergency situations. He described IK as focusing on mobilizing societies to deal with change, yet colonialism deliberately dismantled the kinship qualities in indigenous societies that foster this coordination. As a result, he believes that uplifting IK involves rebuilding these qualities and increasing coordination. Whyte provided an example of how IK can inform the discussion of climate tipping points. He noted that people often use the term *hope* when they discuss whether the United States will significantly address climate change, but that Native peoples regard hope as an emotion that is invoked once you have already passed a tipping point. He pointed out that from an indigenous perspective, hope is not necessary if action is taken before a tipping point has been tipped, when it is known what needs to be done to manage risks and mobilize people.

Whyte ended his presentation by suggesting that IK has been missing from the discussion of climate change for too long, and that nations, nonprofits, and corporations may not have the ability to summon the levels of coordination needed to avert big crises. He noted that because of the longtime absence of IK in discussions regarding global change, global society may be past the point of being able to possess the necessary political and social relationships to curb climate change.

CASCADING IMPACTS AND ENVIRONMENTAL JUSTICE

Dr. Michael Mendez, University of California, Irvine, focused his remarks on several case studies of tipping points oriented around low-income communities of color in the United States. Echoing Dr. Whyte's remarks, Mendez argued that from an environmental justice perspective, climate impact and climate-exacerbated hazard tipping points have already long passed for communities of color and that the lack of coordinated responses may be driven by the disproportionately underprivileged status of the people affected. Mendez emphasized that climate impacts are not equally distributed among the population—the most socially vulnerable are bearing the brunt of the impacts (Figure 4-1). He noted that these impacts are expected to become more frequent and severe over time, leading to more air and water pollution, severe wildfires, heatwaves, and other hazards. Mendez explained that compounding climate impacts and subsequent heightened disaster exposure in places such as California, Texas, and Florida exacerbate existing disparities. Mendez stressed the urgency and importance of understanding how these events and impacts amplify preexisting inequalities and how to lessen the harms.

Mendez stated that climate events, such as extreme heat in California, tend to compound other hazards and comorbidities. Mendez explained that Latino and indigenous migrant communities, which already experience marginalization, disenfranchisement, and discrimination, endure hyper marginalization when climate and compounding events occur. He cited the Centers for Disease Control and Prevention, which reported that heat deaths in migrant workers are 20 times greater than those among U.S. civilian workers.[1] He noted that these higher rates are likely the result of heightened vulnerability during harvesting months, which coincides with frequent extreme weather and disaster events such as heat and wildfires.

---

[1] See https://www.cdc.gov/mmwr/preview/mmwrhtml/mm5724a1.htm.

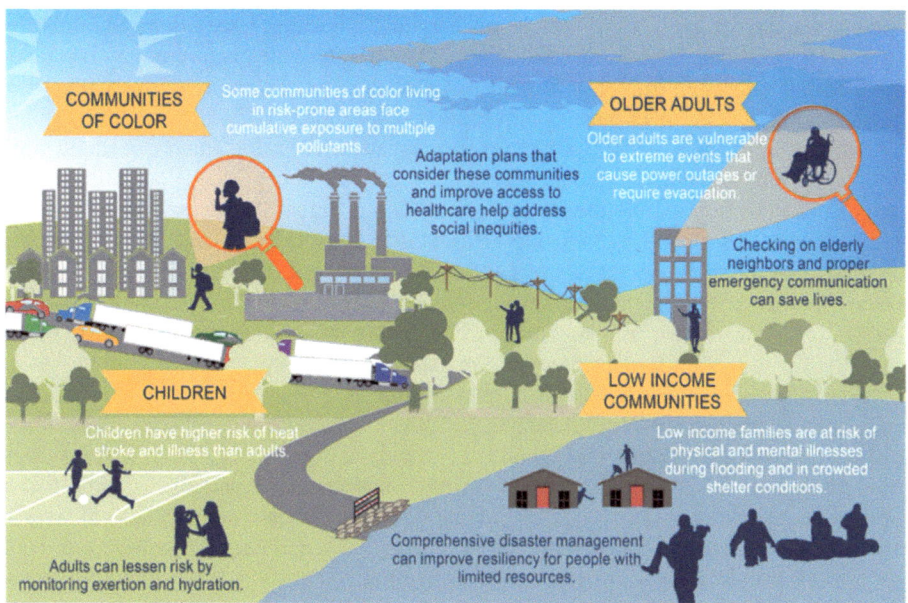

FIGURE 4-1 Demographics of who are most vulnerable to extreme heat.
SOURCE: Fourth National Climate Assessment, Chapter 14, Figure 14.2 (USGRP, 2018).

Mendez explained that nongovernmental organizations, such as indigenous migrant rights organizations, were originally established to help immigrants maintain labor rights in the face of compounding risks and vulnerability. However, these organizations are now advocating for and assisting with climate and environmental justice issues, he noted, because of insufficient planning and response from federal, state, and local governments for undocumented immigrants. Mendez described how these organizations have been providing assistance and relief to these communities on insufficient budgets, with organizational structures that were never designed with the capacity to address these issues. As a result, Mendez noted, they are often unable to address all the needs of these marginalized communities on their own.

Mendez then described an event convened by organizations that provide monetary, emergency, and health care relief to undocumented immigrant communities impacted by climate change. He explained that the goal for the Undocufund Summit,[2] held in September 2022, was to ensure that the current lack of coordination and funding around these issues does not continue. He emphasized that even in areas that are considered progressive, such as California, more work is needed to resolve critical issues, such as disaster planning, language issues, justice, and cultural sensitivity. Mendez noted that issues of language were

---

[2] See https://undocufund.org/2022/09/16/the-first-ever-undocufund-summit-hosted-by-805-undocufund-undocufund-and-latino-community-foundation-calling-for-equitable-disaster-relief/.

particularly acute for speakers of indigenous languages, because translations and interpreters are rarely provided. Mendez argued that the lack of available resources may be due not only to social vulnerabilities, but also to political choices to intentionally withhold resources and knowledge from these communities.

He compared the average costs incurred by an attendee to the famed Coachella music festival (~$2,500) to the income of an average farm laborer in the Coachella Valley (annual salary of $18,000) to illustrate the inequities between different demographic groups in the region. Mendez described how housing affordability is a major issue in the region, with many undocumented immigrant laborers living in cars or sleeping on the ground to combat persistent extreme heat at night. He further noted that Southern California is the only place in the United States where heat deaths have been documented in winter, as a result of hot, dry winds. He mentioned the Intergovernmental Panel on Climate Change Sixth Assessment Report (IPCC, 2021), which highlights the role of anthropogenic climate change in intensifying weather temperature extremes that may disproportionately affect communities of color in urban areas because of urban heat islands and increased smog. Mendez highlighted a new project[3] in which he is involved that measures the urban heat island effect on majority-Latino schools in Los Angeles and noted that lack of greenery and natural surfaces contributed to extremely high temperatures on playground surfaces. He said that some schools in the project have as little as 3 percent to 4 percent green space on their properties, which intensifies extreme heat effects and increases children's vulnerability to heat.

Mendez ended his presentation by highlighting the low diversity in persons working in disaster management, which indicates a lack of representation of communities that are hardest hit by climate change impacts. He emphasized a need to engage migrant communities and advocacy groups to better protect these communities and ensure better outcomes.

MODERATED Q&A DISCUSSION WITH EXPERTS

Presentations by Whyte and Mendez were followed by participant questions and discussion. The sections below summarize specific questions asked to each speaker, followed by connecting themes.

Discussion on Indigenous Knowledge

Following his presentation, Whyte was asked to elaborate on the best ways to connect with indigenous people to facilitate co-production of knowledge and work toward positive social tipping points for climate. He noted the importance of spending proper time to build trusting relationships and stressed how rushing co-production runs the risk of violating trust and consent. He emphasized that from an indigenous lens, climate tipping points have already passed and can no longer be averted.

Regarding IK of biodiversity compared to climate change, Whyte suggested that at the international level there is more collaboration and respect for indigenous land stewards

---

[3] Alliance for a Better Community is a Los Angeles–based Latino/a advocacy nonprofit organization (see https://afabc.org/).

but that this integration does not extend to international cooperative units on climate change. He highlighted that Native advocates who had been working in other areas have added climate to their portfolios but that the visibility of this newer work remains insufficient.

Asked to comment on kinship reciprocity with the nonhuman or more-than-human world, Whyte acknowledged that occasionally scientists are skeptical of this type of kinship relationship. He emphasized that skepticism interferes with a person's ability to steward their environment and violates the empirical nature of IK. Whyte described differences in kinship relationships between people and nonhuman entities in terms of the ability to communicate. In relationships with humans with whom you share a common language, he noted that having respect for consent is relatively easy to understand, but it is more complex when thinking about impacts on animals or other nonhuman entities.

Discussion on Environmental Justice

During the discussion following his presentation, Mendez elaborated on how activities such as the Undocufund Summit could increase awareness and recognition of issues of civil rights and climate adaptation justice for vulnerable communities. He emphasized that environmental issues are now driving a great deal of migration and that migrants fleeing challenging climate situations in their home countries are themselves the most vulnerable to climate issues in the United States. He explained that the Summit was designed to acknowledge these common themes across immigrant groups and regions and to strategize for more state resources for vulnerable communities to combat disasters such wildfires, flooding, and extreme heat. He noted that some action toward this goal is already under way but that more is needed.

In response to a question from a participant, Mendez suggested that the response of the U.S. Environmental Protection Agency's National Environmental Justice Advisory Council Program to the impacts of heatwaves on Latino and indigenous communities in California represented a good starting point. He mentioned that one of the administrators had extensive experience with regional environmental justice organizations and expressed hope that they would be able to upscale lessons from the state level to the national level.

During the discussion, Mendez described needed actions related to energy equity and resilience among highly exposed populations. He said that it was critical to acknowledge and include community experience, as they are quickly developing expertise in disaster resilience due to their exposure and unmet needs. Mendez emphasized that in many places in the United States, extreme events are no longer "unpredictable" and stressed the importance of integrating vulnerable populations and organizations as trusted partners. In particular, he highlighted the need for risk communication and interpretation in Spanish and Indigenous languages.

Asked to comment on the economic impacts of extreme heat in California and how communities can help define state priorities, Mendez highlighted challenges with physical infrastructure but also noted that underserved communities have poor digital infrastructure to support online and cellular communications. He stressed the importance of these communication mechanisms for getting emergency information, particularly in the context of adjusting power usage to help prevent blackouts during extreme heatwaves. He added that loss of life and poor health have serious economic consequences in terms of lost productivity.

Cross-Cutting Themes

The two discussions following Whyte's and Mendez's presentations highlighted a few cross-cutting themes:

1. **Community Engagement and Co-Production of Knowledge:** Both Whyte and Mendez emphasized the importance of building trust and long-term relationships with indigenous communities and communities of color to facilitate co-production of knowledge related to climate change impacts and adaptation strategies. Whyte stressed that building trust involves respecting indigenous knowledge systems and understanding that meaningful collaboration takes time.
2. **Recognition of Systemic Inequities:** Both speakers mentioned a focus on acknowledging systemic injustices and disparities, particularly concerning vulnerable communities such as undocumented immigrants and low-income populations. Mendez noted that climate change exacerbates existing inequalities and that addressing these issues requires inclusive and equitable approaches to policymaking and resource allocation, while Whyte emphasized sharing of knowledge.
3. **Integration of Environmental and Social Justice Movements:** Efforts to address climate change impacts need to be integrated with broader social justice movements, including labor rights advocacy, immigration reform, and educational equity, Mendez highlighted. Both Whyte and Mendez discussed how building intersectional coalitions can amplify the voices of marginalized communities, knowledge, and perspectives, and drive meaningful policy and institutional changes.
4. **Cascading Impacts and Multidimensional Risks:** Both speakers discussed how the cascading impacts of climate change cause societal disruptions and emphasized that understanding these interconnected risks is fundamental to understanding climate change impacts and to developing comprehensive adaptation and resilience strategies.
5. **Language Justice and Cultural Competence:** Mendez reiterated that language justice is a critical issue to ensure that emergency information and communication efforts are linguistically accessible to all communities. Both Whyte and Mendez stressed the importance of cultural competence and sensitivity for effective communication around climate impacts and policy implementation.

BREAKOUT GROUP DISCUSSIONS ON INTERDISCIPLINARY RESEARCH TOPICS AND OPPORTUNITIES

During the final afternoon of the workshop, the breakout discussions were framed around three themes:

1. Developing equitable research strategies
2. Developing new research capabilities
3. Strategies to advance transdisciplinary research

All participants, both in-person and virtual, had the opportunity to participate in discussions on all three topics. A rapporteur from each group was assigned to share key themes from their group's discussion. The recurring themes from these discussions are summarized in the sections below.

Developing Equitable Research Strategies

Dr. Simon Dietz, London School of Economics and Political Science, summarized the recurring themes from breakout group discussions on developing equitable research strategies. He explained that diversity is a complex, multidimensional concept and emphasized the importance of coming to a shared understanding about what kinds of diversity to target. Dietz highlighted discussion on the importance of long-term relationship and trust building, which involves funding and institutional support. He also noted the importance of co-production from the beginning, with equal consideration given to different knowledge types and complementary opportunities for learning across types. National Academies of Sciences, Engineering, and Medicine staff member, Hugh Walpole, echoed that the virtual group discussion followed the theme of bringing in diverse and affected people as participants in research, with a strong group focus on grassroots organizations. He also described interest in integrating technology with indigenous knowledge. Dietz detailed related discussion on the importance of more diverse leadership, including diversity in age, to push for equity and inclusion across different organizational levels. Finally, he highlighted comments about enhancing young researchers' skills in communicating research and participating in transdisciplinary research.

Developing New Research Capabilities

Dr. Dorothy Merritts, Franklin and Marshall College, summarized the breakout discussion on developing new research capabilities. She explained that the groups considered new computational, experimental, and organizational capabilities that could advance transformation in understanding and predicting tipping points, as well as their cascading impacts and interactive risks. She highlighted that all three groups elevated the importance of communication, including using storylines or evidence-based narratives to foster communication across social and physical disciplines and diverse groups. Other ideas she shared were the exercises of anticipatory thinking, highlighting how a participant shared that some indigenous cultures think seven generations into the future, and "what-if?" scenarios, which are highly effective for thinking about tipping points. Merritts also highlighted a discussion theme on advances in remote sensing that could be used to gain insights into resilience, such as the potential to track communities during disasters and recovery. She noted discussion about the use of artificial intelligence to inform understanding of social and physical sciences and their interconnectedness. Walpole added that virtual participants discussed big data and the importance of improving modeling, with a particular focus on the need for immediate action to address climate change.

Strategies to Advance Transdisciplinary Research

Dr. Amir AghaKouchak, University of California, Irvine, led discussions on advancing transdisciplinary research. He described how discussions considered how advances across the natural, social, computational, and engineering sciences could be integrated to construct practical understanding. AghaKouchak highlighted that a key discussion theme was community-focused, co-produced, bottom-up research strategies to promote transdisciplinary research in broad, solution-oriented research questions. As an example, he described the American Geophysical Union's Thriving Earth Exchange, which aims to bring together diverse researchers and disciplines to address solution-oriented challenges. Echoing themes relayed by Merritts in the previous report-back, he noted that many participants expressed interest in artificial intelligence and remote sensing as technologies that can inform science across disciplines and foster collaboration.

He also described participant discussion around increasing the number of practical solutions by bringing experts in governance into interdisciplinary research, considering expert judgment in parallel with technical modeling, fostering receptiveness to indigenous knowledge, and integrating behavioral responses and human dimensions into climate models. He detailed group conversations about framing solutions around urgency and desirable solutions rather than existing circumstances. He closed by highlighting discussion around utilizing interdisciplinary assessment of research to help understand key needs and motivate collaboration. Walpole highlighted that virtual participants discussed accessibility and incentives, including the importance of accessible forums for information and incentives for building multidisciplinary collaboration across disciplinary boundaries.

5
Workshop Synthesis, Themes, and Closing Thoughts

SYNTHESIS OF WORKSHOP DISCUSSIONS

Kristen St. John, James Madison University, provided a summary of the workshop presentations and discussions. She shared that the research directions on tipping points, cascading impacts, and interacting risks that were presented and discussed aligned with the five aspects of the vision for the *Next Generation Earth Systems Science at the National Science Foundation* (NASEM, 2022) consensus report:

- Taking integrated approaches for research on tipping points, impacts, and associated risks
- Giving equal importance to natural and social systems/processes
- Exploring essential interconnections and feedbacks
- Working across a wide range of scales
- Valuing diversity, equity, and inclusion in the full research process

St. John stated that the workshop opened with a discussion of the definitions for tipping points. She noted that operational definitions for tipping points span many different scales, from fundamental global physical elements to extremely localized social impacts. She highlighted that considering tipping points as crossing a threshold, followed by abrupt change, was a general theme throughout the workshop discussions. The concept of reinforcing feedback mechanisms was not as common a thread across discussions, St. John added, despite the importance of this concept to both the social and natural science literatures around tipping points. St. John also highlighted that abrupt change means very different things at different temporal scales, creating challenges for thinking about scaling in terms of tipping points, and that some terminology around tipping points (such as "bomb cyclone" and "atmospheric rivers") have implications for public understanding. She stressed that regionally focused narratives have the potential to add tangible meaning to discussions around cascading impacts and interacting risks and noted that this may be a critical approach for communicating about these issues and facilitating effective decision-making.

St. John then highlighted some "general rules." First, she pointed out that societal tipping points tend to occur faster than physical tipping points, while ecological tipping points are intermediate in time. Second, she explained that physical and chemical processes tend to have more linear relationships, but biological physiological rate laws tend to build more complex relationships. St. John added that these more complex, parabolic relationships are useful for understanding the complexities of the system and how they give rise to alternative stable states, bridged by a tipping point.

HIGH-LEVEL WORKSHOP THEMES

St. John framed her following discussion around the high-level themes that emerged from the workshop presentations and discussions. She explained that these themes (described below) are organized around questions derived from the workshop Statement of Task.

Theme 1: Lessons from Historical Analysis

*What can be learned from **historical analysis** of past physical and social tipping points to inform understanding, prediction, and preparation in the future?*

St. John highlighted a takeaway theme from workshop discussions that conducting tipping points research at and across different scales is critical, because different scales provide useful lenses for understanding the impacts of people on climate. She shared examples from both ends of the temporal scale spectrum, including using the paleoclimate record to identify the state of the climate prior to human civilization and historical records from ancient to recent history. St. John noted that these lenses provide critical insights for understanding risks to the human system. She highlighted that several participants wondered whether the dynamics of social systems have changed fundamentally in the relatively recent past, creating challenges for learning from recent history.

Theme 2: Outstanding Research Questions

*What are the **key outstanding research questions** on physical and social tipping points, the interacting risks of these tipping points, and their cascading impacts?*

St. John emphasized a series of questions and themes related to these points. The first point highlighted the benefits of integrating natural and social process research with the understanding that physical tipping points will likely have a variety of currently unknown social impacts. She stressed the need to better understand, measure, model, and visualize the coupling between nonlinear physical changes and subsequent impacts on humans and society.

She discussed the need to better understand interactions between tipping points and their cascading impacts across scale. She noted the importance of studying the teleconnections of physical changes across spatial and timescales, where tipping points in a particular region or point in time contribute as inputs to tipping points in other regions at other times. She emphasized discussions around the work still needed to learn more about physical tipping points themselves.

On social tipping points, St. John highlighted several areas where current understanding is incomplete, such as how to account for agency, adaptation of social systems, and potential mitigation in social tipping points. She pointed out the large quantity of relevant research across economic and social sciences but noted that the lens of social tipping points was rarely used to frame the results of that research. St. John highlighted a discussion theme around early warning signals for tipping points. She explained that several generic signs across tipping elements act as early warning signals that a tipping point is approaching. However, she also noted open questions about how to apply those warning signals to regional and local scales, where most decisions that govern services to address outcomes (e.g., medical and social services) will be made.

Finally, St. John emphasized discussions on acceptable levels of risk and uncertainty when conducting tipping points research.

Theme 3: Barriers and Opportunities for Progress

*What are major **barriers** and **opportunities** to accelerate progress to advance these areas of research?*

Regarding this theme, St. John listed seven barriers that were highlighted during the workshop:

- Challenges working across scales of time, space, and social organization
- Challenges in forming and organizing transdisciplinary research teams
- Tension between economic goals and those of Earth systems in support of human well-being
- Barriers in implementation of research outcomes, especially where there is a range of ability and/or agency for individuals to address risks and outcomes from climate change
- Lack of standards and baselines for important concepts such as resilience
- Methodological gaps between different disciplines/scales of study
- History of inequity and exclusion of underrepresented groups (e.g., people of color, indigenous communities, low-income communities) from scientific discourse

St. John also highlighted five opportunities for transdisciplinary research discussed during the workshop:

- Quantifying the economic and social costs of extreme weather events can provide a useful mechanism for communicating about the outcomes of climate tipping points.
- Social responses to extreme weather events can serve as useful sources of evidence for the dynamics of social climate tipping points.
- Designing climate adaptation options provides opportunities to make design choices to make progress toward environmental justice goals simultaneously.

- Enthusiasm in the research community (and workshop participants) for transdisciplinary research
- Growing awareness of the benefits and importance of engaging with indigenous stakeholders and knowledge producers at all stages of the research process

Theme 4: Diverse, Equitable, and Inclusive Tipping Points Research

*How can tipping point research priorities be identified, studies be conducted, and response strategies be designed in a manner that is **inclusive and equitable for a diversity of participants and engaged stakeholders**? What additional **perspectives** and considerations could be incorporated into the understanding of interaction risks?*

St. John emphasized several themes centered on inclusion and equity. She highlighted discussion about including from the initial stages those who are most vulnerable to the outcomes and consequences of climate tipping points in the research process. She noted discussions about indigenous perspectives and the science of coordination to apply knowledge gained from research and to support intersectional coalitions. She pointed out the importance of combining funding streams from government research and non-governmental trusts and foundations to diversify the perspectives being addressed in the research space.

Theme 5: Building Integrated and Practical Understanding

*How can advances across the natural, social, computational, and engineering sciences **be integrated** to **build practical understanding** of tipping points, cascading impacts, and interacting risks?*

St. John highlighted discussion around several key points, including providing incentives for transdisciplinary research in circumstances where such research practices would be beneficial. She underscored the importance of engagement with non-academic and non-scientific audiences early in the process, as well as engaging in co-production to build practical understanding around tipping point issues as a mechanism for creating change. She noted that there was a diverse range of perspectives included in the workshop, but that most were from academia, and underscored that broader outreach outside of academic audiences is important for practical understanding. St. John also highlighted the importance of monitoring and assessment mechanisms for research, particularly for small but scalable research projects that could be adapted at broader scales. St. John added how important communications research is to contribute to the practical understanding of critical elements to tipping points research and engaging with stakeholders and decision-makers in the broadest, most effective ways.

Theme 6: New Capabilities for Transformational Advances

*What **new capabilities**—computational, experimental, organizational, etc.—would lead to transformational advances in the understanding and prediction of tipping points, cascading impacts and interacting risks?*

St. John listed four areas that emerged from workshop discussions where new capabilities could be helpful for tipping points: organizational, conceptual, computational, and observational.

Under organizational capabilities, St. John highlighted the need for approaches to efficiently connect people who do not typically communicate with one another. She noted that planning grants at multi-institutional levels can help achieve this.

For conceptual capabilities, St. John provided the example of systems visualizations to help conceptually link social and natural processes. She highlighted the example of recent IPCC reports using storylines to communicate about connected processes and suggested that such connections can impact both how research is conducted and how it is communicated.

For computational capabilities, St. John noted the potential of machine learning and artificial intelligence techniques to aid in modeling of tipping points and to support early warning capabilities.

For observational capabilities, St. John noted three key areas—advancing the use of existing remote sensing capabilities, deploying additional remote sensing capabilities where necessary, and developing new observable proxy factors to stand in for other challenging-to-measure properties of tipping systems.

CLOSING THOUGHTS

St. John ended her remarks by highlighting possible framings and opportunities to propel future research. She emphasized the importance of designing strategies that enhance transdisciplinary connections, using social science to help build successful transdisciplinary research teams, and expanding the inclusion of historically marginalized stakeholders and groups. She noted the potential impact of elevating discussions around regionally scaled tipping points, with opportunities that focus on regional boundary conditions and the risks and impacts associated with regional tipping factors. St. John also highlighted earlier discussions on communication on positive tipping points and stressed the potential to manage the risks around negative tipping points in a more positive framing. St. John encouraged exploration of the opportunities or risks presented by triggering positive tipping points. Finally, she highlighted two potential adjustments in participants' thinking: (1) broadening the vision of society to identify potential change agents within existing systems, including drawing on social science and indigenous knowledge and (2) focusing on the most vulnerable regions, ecosystems, and communities for prioritizing tipping point research while documenting outcomes and inequities in those systems.

References

Allen, Craig R. 2023. "Tipping Points and Cascading Impact in the Great Plains, USA." Presented on January 18, 2023. Workshop on Tipping Points, Cascading Impacts, and Interacting Risks in the Earth System, National Academy of Sciences, Washington, D.C.

Armstrong McKay, David I., Arie Stahl, Jessee F. Abrams, et al. 2022. "Exceeding 1.5°C Global Warming Could Trigger Multiple Climate Tipping Points. *Science* 377:eabn 7950. https://doi.org/10.1126/science.abn7950.

Brovkin, Victor, Edward Brook, John W. Williams, et al. 2021. "Past Abrupt Changes, Tipping Points and Cascading Impacts in the Earth System." *Nature Geoscience* 14:550–8. https://doi.org/10.1038/s41561-021-00790-5.

Chapin, F. Stuart, Elke U. Weber, Elena M. Bennett, et al. 2022. "Earth Stewardship: Shaping a Sustainable Future through Interacting Policy and Norm Shifts." *Ambio*. https://doi.org/10.1007/s13280-022-01721-3.

Formanski, Felix J., Marcel M. Pein, David D. Loschelder, et al. 2022. "Tipping Points Ahead? How Laypeople Respond to Linear Versus Nonlinear Climate Change Predictions." *Climatic Change* 175:8. https://doi.org/10.1007/s10584-022-03459-z.

Fu, Lee-Lueng, Edward J. Christensen, Charles A. Yamarone Jr., et al. 1994. "TOPEX/POSEIDON Mission Overview." *Journal of Geophysical Research* 99:C12. https://doi.org//10.1029/94JC01761.

Gaffney, Owen, Zoe Tcholak-Antitch, Sophia Boehm, et al. 2021. *Global Commons Survey: Attitudes to Planetary Stewardship and Transformation among G20 Countries.* Global Commons Alliance. https://globalcommonsalliance.org/wp-content/uploads/2024/01/Global-Commons-G20-Survey-full-report.pdf.

Gladwell, Malcolm. 2000. *The Tipping Point: How Little Things Can Make a Big Difference.* Boston: Little, Brown and Company.

Grant, Gordon E. 2023. "Summer Water in the West: Climate, Tipping Points, and Cascading Impacts." Presented on January 18, 2023. Workshop on Tipping Points, Cascading Impacts, and Interacting Risks in the Earth System, National Academy of Sciences, Washington, D.C.

Hall, Kara L., Amanda L. Vogel, Brooke A. Stipelman, et al. 2012. "A Four-Phase Model of Transdisciplinary Team-Based Research: Goals, Team Process, and Strategies." *Translational Behavioral Medicine* 2(4):415–30. https://doi.org/10.1007/s13142-012-0167-y.

Hart, Julian T. 1971. "The Inverse Care Law." *The Lancet* 297(7696):405-12. https://doi.org/10.1016/S0140-6736(71)92410-X.

References

Heger, Martin P., and Eric Neumayer. 2019. "The Impact of the Indian Ocean Tsunami on Aceh's Long-Term Economic Growth." *Journal of Development Economics* 141:102365. https://doi.org/10.1016/j.jdeveco.2019.06.008.

IPCC (Intergovernmental Panel on Climate Change). 2023. *Climate Change 2021: The Physical Science Basis. Contribution of Working Group I to the Sixth Assessment Report of the Intergovernmental Panel on Climate Change,* edited by V. Masson-Delmotte, P. Zhai, A. Pirani, et al. Cambridge: Cambridge University Press. doi:10.1017/9781009157896.

Kump, Lee R., and James E. Lovelock. 1995. "The Geophysiology of Climate." Chapter 15 in *World Survey of Climatology* (vol. 16), edited by A. Henderson-Sellers, 537–53. Elsevier. https://doi.org/10.1016/S0168-6321(06)80038-6.

Lenton, Timothy M., Johan Rockström, Owen Gaffney, et al. 2019. "Climate Tipping Points—Too Risky to Bet Against." *Nature*. https://www.nature.com/articles/d41586-019-03595-0.

Lenton, Timothy M. 2020. "Tipping Positive Change." *Philosophical Transactions of the Royal Society*. http://doi.org/10.1098/rstb.2019.0123.

Lovelock, James E. 1972. "Gaia as Seen through the Atmosphere." *Atmospheric Environment* 6:579-580. https://doi.org/10.1016/0004-6981(72)90076-5.

Mack, Michelle. 2023. "Impacts of Arctic and Boreal Elements at Local and Global scales." Presented on January 18, 2023. Workshop on Tipping Points, Cascading Impacts, and Interacting Risks in the Earth System, National Academy of Sciences, Washington, D.C.

McCabe, Tempest D., and M.C. Dietze. 2019. "Scaling Contagious Disturbance: A Spatially-Implicit Dynamic Model." *Frontiers in Ecology and Evolution* 7. https://doi.org/10.3389/fevo.2019.00064.

Margulis, Lynn and James E. Lovelock. 1973. "Biological modulation of the Earth's atmosphere." *Icarus* 21(4):471-489. https://doi.org/10.1016/0019-1035(74)90150-X.

Meadows, Donella. 2015. "Leverage points-places to intervene in a system." Hartland, VT: Sustainability Institute.

Montaggioni, Lucien F. 2005. "History of Indo-Pacific Coral Reef Systems Since the Last Glaciation: Development Patterns and Controlling Factors." *Earth Science Reviews* 71 (1-2): 1–75. https://doi.org/10.1016/j.earscirev.2005.01.002.

Moser, S.C., and Hart, J.A.F. 2015. The long arm of climate change: societal teleconnections and the future of climate change impacts studies. *Climatic Change* 129:13–26. https://doi.org/10.1007/s10584-015-1328-z.

NASEM (National Academies of Sciences, Engineering, and Medicine). 2022. *Next Generation Earth Systems Science at the National Science Foundation.* Washington, DC: The National Academies Press. https://doi.org/10.17226/26042.

Newman, R., and I. Noy. 2023. "The Global Costs of Extreme Weather That Are Attributable to Climate Change." *Nature Communications* 14 (6103). https://doi.org/10.1038/s41467-023-41888-1.

Nigam, Sumant. 2003. *TELECONNECTIONS: Encyclopedia of Atmospheric Sciences* (1st ed.), edited by James Holton, 2243–69. Academic Press. https://doi.org/10.1016/B0-12-227090-8/00400-0.

Noy, Ilan. 2023. "An Economist's View of Cascades and Tipping Points." Presented on January 17, 2023. Workshop on Tipping Points, Cascading Impacts, and Interacting Risks in the Earth System, National Academy of Sciences, Washington, D.C.

Olsson, Per, Carl Folke, and Thomas. Hahn. 2004. "Social-Ecological Transformation for Ecosystem Management: The Development of Adaptive Co-Management of a Wetland Landscape in Southern Sweden. *Ecology and Society* 9(4):2. http://www.ecologyandsociety.org/vol9/iss4/art2/.

Storlazzi, Curt D., Stephen B. Gingerich, Ap van Dongeren, et al. 2018. "Most Atolls Will Be Uninhabitable by the Mid-21st Century Because of Sea-Level Rise Exacerbating Wave-Driven Flooding." *Science Advances* 4. https://doi.org/10.1126/sciadv.aap9741.

Storlazzi, Curt. 2023. "Tipping Points in Future Tropical Pacific Island Sustainability." Presented on January 18, 2023. Workshop on Tipping Points, Cascading Impacts, and Interacting Risks in the Earth System, National Academy of Sciences, Washington, D.C.

United States Global Change Research Program (USGCRP). 2018. "Human Health." Chapter 14 in *Impacts, Risks, and Adaptation in the United States: Fourth National Climate Assessment, Volume II*, edited by D.R. Reidmiller, C.W. Avery, D.R. Easterling, et al. U.S. Global Change Research Program, Washington, DC, 1515 pp. https://doi.org/10.7930/NCA4.2018.

Appendix A
Workshop Agenda

COMMITTEE ON TIPPING POINTS, CASCADING IMPACTS, AND INTERACTING RISKS IN THE EARTH SYSTEM

NAS Building, 2101 Constitution Ave NW
Washington, DC 20418

JANUARY 17, 2023
Virtual
OPENING SESSION

10:00–10:10	Welcome and Housekeeping **Kristen St. John,** *Chair* **Margo D. Corum,** National Academies of Sciences, Engineering, and Medicine **Patricia Razafindrambinina,** National Academies of Sciences, Engineering, and Medicine **Anjuli Bamzai,** National Science Foundation **Candace Major,** National Science Foundation
10:10–10:40	Background, Introduction, Pre-Workshop Poll **Kristen St. John,** *Chair*
10:40–11:40	Overview on Tipping Points and Expert Response **Dorothy J. Merritts,** *Moderator* **Timothy Lenton,** *Speaker*, Director and Chair, Global Systems Institute and Climate Change and Earth System Science at the University of Exeter **Jeffrey N. Rubin,** *Speaker* **Robert Kopp,** *Speaker*, Professor and Co-director, University Office of Climate Action at Rutgers University
11:40–12:15	Break

SESSION 1
Historical Analysis of Past Biogeophysical and Social Tipping Points

12:15–1:15　　**Michael Schoon,** *Moderator*
Ilan Noy, *Speaker*, Chair, Economics of Disasters and Climate Change at Victoria University of Wellington
Benjamin I. Cook, *Speaker*, Research Physical Scientist, National Aeronautics and Space Administration Goddard Institute for Space Studies
Lee Kump, *Speaker*, Professor, Pennsylvania State University
(**Timothy Lenton,** *Speaker*, Director and Chair, Global Systems Institute and Climate Change and Earth System Science at the University of Exeter, presented Kump's remarks on his behalf)

1:15–2:00　　Breakout Discussions and Report Out
1. What are the key outstanding questions on biogeophysical and social tipping points, the interacting risks of these tipping points, and their cascading impacts?
2. What are the major barriers and opportunities to accelerate progress to advance these areas of research?

2:00　　END OF DAY 1

JANUARY 18, 2023
Virtual
OPENING SESSION

10:00–10:10　　Welcome and Housekeeping
Kristen St. John, *Chair*
Margo D. Corum, National Academies of Sciences, Engineering, and Medicine
Patricia Razafindrambinina, National Academies of Sciences, Engineering, and Medicine

SESSION 2
Regional Perspectives on Climate Tipping Points and Cascading Impacts

10:10–11:30　　**Simon Dietz,** *Moderator*
Curt Storlazzi, *Speaker*, Senior Research Geologist, U.S. Geological Survey
Gordon Grant, *Speaker*, Research Hydrologist, U.S. Department of Agriculture Forest Service

Appendix A

11:30–12:00	Break
12:00–1:20	**Dorothy J. Merritts,** *Moderator* **Terry Chapin,** *Speaker*, Professor Emeritus, University of Alaska **Michelle Mack,** *Speaker*, Professor, Northern Arizona University **Craig Allen,** *Speaker*, Professor, University of Nebraska–Lincoln
1:20–2:00	Breakout Discussions and Report Out 1. What are the tipping points, cascading effects, and interacting risks critical to these geographic regions? (Coastal U.S., American West, Arctic, and Great Plains) 2. Identify transdisciplinary research opportunities.
2:00	END OF DAY 2

JANUARY 19, 2023
Virtual
OPENING SESSION

10:00–10:05	Welcome and Housekeeping **Kristen St. John,** *Chair* **Margo D. Corum,** National Academies of Sciences, Engineering, and Medicine **Patricia Razafindrambinina,** National Academies of Sciences, Engineering, and Medicine

SESSION 3
Interdisciplinary Research Priorities and Opportunities

10:05–11:05	**Amir AghaKouchak,** *Moderator* **Kyle Whyte,** *Speaker*, Willis Pack Professor, University of Michigan **Michael Méndez,** *Speaker*, Assistant Professor, University of California, Irvine
11:05–11:30	Break
11:30–1:00	Breakout Discussions and Report Out 1. How can research priorities be identified, studies be conducted, and response strategies be designed in a manner that is inclusive and equitable for a diversity of participants and engaged stakeholders? 2. What new capabilities–computational, experimental, and organizational, etc.–would lead to transformational advances in the

understanding and prediction of tipping points, cascading impacts, and interacting risks?
3. How can advances across the natural, social, computational, and engineering sciences be integrated to build practical understanding?

CLOSING REMARKS
Summary, Synthesis, and Next Steps

1:00–1:45   **Kristen St. John,** *Chair*

1:45   Meeting Adjourns

Appendix B
Statement of Task

A planning committee of the National Academies of Sciences, Engineering, and Medicine will plan a workshop[1] to bring together experts to consider tipping points, cascading impacts, and interacting risks in the Earth system, incorporating the key characteristics laid out in the National Academies report, *Next Generation Earth Systems Science at the National Science Foundation* (NASEM, 2022). The workshop will apply Earth systems science approaches to explore understanding, prediction, and preparation for tipping points in the Earth system and cultivate connections among the transdisciplinary research community. The presentations and discussions at the workshop will be published as workshop proceedings prepared by rapporteur in accordance with institutional guidelines. Specific questions to be discussed may include:

- What are the key outstanding research questions on physical and social tipping points, the interacting risks of these tipping points, and their cascading impacts?
- What are the major barriers and opportunities to accelerate progress to advance these areas of research?
- What can be learned from historical analysis of past physical and social tipping points to inform understanding, prediction, and preparation in the future?
- How can advances across the natural, social, computational, and engineering sciences be integrated to build practical understanding?
- What new capabilities—computational, experimental, organizational, etc.—would lead to transformational advances in the understanding and prediction of tipping points, cascading impacts, and interacting risks?
- How can research priorities be identified, studies be conducted, and response strategies be designed in a manner that is inclusive and equitable for a diversity of participants and engaged stakeholders? What additional perspectives and considerations could be incorporated into the understanding of interacting risks?

---

[1] See https://www.nationalacademies.org/our-work/workshop-on-tipping-points-cascading-impacts-and-interacting-risks-in-the-earth-system.

Appendix C
Planning Committee Biographies

Kristen St. John (*Chair*) is a Professor of Geology at James Madison University. She earned her M.S. and Ph.D. in Geoscience from The Ohio State University. Her research focuses on marine sediment records of past climate change, and on teaching and learning in the geosciences. An active researcher in the International Ocean Discovery Program (IODP), she was a marine sedimentologist for several expeditions, and worked on samples from the Arctic, North Atlantic, North Pacific, and South Pacific. Dr. St. John is the U.S. co-chief scientist for IODP Expedition 377: Arctic Ocean Paleoceanography (ArcOP; awaiting rescheduling). Her research informs the design and content of the book *Reconstructing Earth's Climate History: Inquiry-Based Exercises for the Lab and Class*. Dr. St. John was the editor-in-chief of the *Journal of Geoscience Education* from 2012 to 2017. She is the President-elect of the American Geophysical Union Education Section and is a Geological Society of America Fellow. Dr. St. John recently served on the NASEM Committee on Advancing a Systems Approach to Studying the Earth: A Strategy for the National Science Foundation, and she serves on the Polar Research Board.

Amir AghaKouchak is a Professor of Civil and Environmental Engineering and Earth System Science at the University of California, Irvine. His research focuses on natural hazards and climate extremes and crosses the boundaries between hydrology, climatology, and remote sensing. One of his main research areas is studying and understanding the interactions between different types of climatic and nonclimatic hazards, including compound and cascading events. He has received several honors and awards, including the American Geophysical Union's James B. Macelwane Medal and the American Society of Civil Engineers (ASCE) Huber Research Prize. Dr. AghaKouchak is currently serving as the editor-in-chief of *Earth's Future*—a transdisciplinary scientific journal examining the state of the planet and the science of the Anthropocene. Dr. AghaKouchak has published 200 peer-reviewed research articles in scientific journals. He has served as the principal investigator of many interdisciplinary research grants funded by the National Aeronautics and Space Administration (NASA), National Science Foundation (NSF), and National Oceanic and Atmospheric Administration (NOAA). Dr. AghaKouchak obtained his B.S.c (2003) and M.S.c (2005) in Civil Engineering from K.N. Toonsi University of Technology in Tehran, Iran. Subsequently, he received his Ph.D. in Civil and Environmental Engineering from the University of Stuttgart, Germany, 2010.

Katharine Cashman (NAS) is a Research Professor at the University of Oregon in the Department of Earth Sciences. She has also held professional positions that include an

AXA Endowed Chair of Volcanology (2014–2021), AXA Research Chair (at the University of Bristol, UK (2011–20140) and the Philip H. Knight Professor of Natural Sciences, University of Oregon (2007–2011). Dr. Cashman is a volcanologist who studies links between chemical and physical factors that control magma ascent, eruption, and emplacement on the Earth's surface. She has studied volcanoes on all seven continents, explored a wide range of eruption styles and has worked on problems that span from the chemical to the physical to social aspects of volcanism. Additionally, Dr. Cashman has worked at the U.S. Geological Survey (USGS) volcano observatories in Hawaii and Washington and served on the Scientific Advisory Committee for the island of Montserrat during the eruption of the Soufriere Hills volcano. Dr. Cashman is a Fellow of the American Geophysical Union (AGU), the American Academy of Arts and Sciences, and the Royal Society (U.K.) and has been elected to the Academia Europaea and the National Academy of Sciences. Dr. Cashman has received the Murchison Medal (Geological Society of London), a Royal Society Wolfson Award, an Honorary Doctorate of Science from Middlebury College and a Bowen Award from the Volcanology, Geochemistry and Petrology section of the AGU. Dr. Cashman is a Distinguished Lecturer for the Mineralogical Society of America and the Mineralogical Society (U.K.) and has given both the Daly lecture and the Bowen lecture at AGU annual meetings, the Shell Lecture at the Geological Society of London, the Inaugural Green Lecture at the University of Leeds and an Annual Address at the Swedish Academy. Dr. Cashman received her Ph.D. in geology from The Johns Hopkins University. She has also served on the NAS Committee on Improving Understanding of Volcanic Eruptions, which produced the consensus report "Volcanic Eruptions and Their Repose, Unrest, Precursors, and Timing."

Simon Dietz is Professor of Environmental Policy at the London School of Economics and Political Science (LSE), where he has a joint appointment between the Grantham Research Institute on Climate Change and the Environment, and the Department of Geography and Environment. He joined the LSE faculty in 2006, was tenured in 2011 and promoted to full professor in 2015. He co-founded the Grantham Research Institute in 2008 and was a director from 2008 to 2017. He is an environmental economist with particular interests in climate change and sustainable development. He has published research on a wide range of issues, including the social cost of carbon, decision-making under uncertainty, and discounting. He has been co-editor of the Journal of the Association of Environmental and Resource Economists since 2016, is a member of the Council of the European Association of Environmental and Resource Economists (and former Vice President), a Center for Economic Studies (CESifo) Research Network Fellow, and a Fellow of the Royal Society of Arts. In 2018, he became the first recipient of the new European Award for Researchers in Environmental Economics under the Age of Forty, "a recognition given every year to the environmental economist under the age of forty who is judged to have made the most significant contribution to environmental economic thought and knowledge." He graduated from the University of East Anglia in 2001 with a B.Sc. (Starred First Class Honours) in Environmental Science, from the LSE in 2002 with an M.Sc. (Distinction) in Human Geography Research, and from the LSE in 2006 with a Ph.D. in Geography and Environment, specializing in environmental economics. Prior to joining the LSE faculty, he was a Policy Analyst at the UK Treasury, where he worked on the Stern Review on the Economics of Climate Change.

Timothy Lenton is founding Director of the Global Systems Institute and Chair in Climate Change and Earth System Science at the University of Exeter. He has more than 25 years, research experience, focused on modelling life's coupling to the Earth system, biogeochemical cycling, climate dynamics, and associated tipping points. His books *Revolutions That Made the Earth* (with Andrew Watson) and *Earth System Science: A Very Short Introduction* have popularized a new scientific view of our planetary home. Tim co-authored the "Planetary Boundaries" framework and is renowned for his work identifying climate tipping points, which won the Times Higher Education Award for Research Project of the Year 2008. He has also received a Philip Leverhulme Prize 2004, European Geosciences Union Outstanding Young Scientist Award 2006, Geological Society of London William Smith Fund 2008, and Royal Society Wolfson Research Merit Award 2013. Tim is a member of the Earth Commission, an Institute for Scientific Information (ISI) Highly Cited Researcher, and in the top 100 of the Reuters "Hot List" of the world's top climate scientists. Tim received his B.S. from Cambridge University in 1994, and obtained his Ph.D. from the University of East Anglia in 1998.

Dorothy Merritts (NAS) is the Harry W. and Mary B. Huffnagle Professor of Geosciences at Franklin and Marshall College. She worked at the U.S. Geological Survey while completing an M.S. in Engineering Geology at Stanford University and a Ph.D. in Geosciences at the University of Arizona. Dr. Merritts's work focuses on the history of landscapes and processes that shape them. She is known particularly for her research on landscapes perturbed by geologic events and climate change during the past ~130,000 years and by human activities during the past ~400 years. Dr. Merritts was president of the American Geophysical Union Earth and Planetary Surface Processes Section, is a fellow of the Geological Society of America (GSA), was a co-recipient of the GSA Kirk Bryan award for outstanding scholarship, received the Distinguished Career award from the GSA Quaternary Geology and Geomorphology Division in 2022, and was elected to the National Academy of Sciences in 2022. Dr. Merritts has served on five National Research Council (NRC) committees, two of which she chaired and one of which she co-chaired, and all of which were related to the impacts of human activities and/or climate change at the Earth's surface.

Michael Schoon is an Associate Professor in the School of Sustainability at Arizona State University, specializing in collaborative environmental governance, the resilience of social-ecological systems, biodiversity conservation, and institutional analysis, taking a transdisciplinary approach to this research. Through his research in social-ecological resilience, Schoon regularly studies and writes about tipping points, transformation, and regime shifts. He is the current chair of the Resilience Alliance, an international, interdisciplinary research organization that focuses on the resilience of social-ecological systems. He is also editor-in-chief of the *International Journal of the Commons*, an academic journal focusing

Appendix C

on the study of common-pool resources and their management. Schoon is active in International Union for Conservation of Nature (IUCN) research in transboundary conservation. His doctoral research won multiple awards, including the American Political Science Association's award for best dissertation. Recently, he convened a collaborative governance group for the US Forest Service to develop recommendations for the management of wild horses in Central Arizona. Schoon received a Ph.D. from Indiana University in an interdisciplinary program in environmental policy, international relations and conservation biology. He worked at Indiana in the Ostrom Workshop, emphasizing governance and institutional analysis.